Ben Stacy Jerrik (Ed.)

Rogóźno, Łask County

AF693896

Ben Stacy Jerrik (Ed.)

Rogóźno, Łask County

Łódź

Part Press

Imprint

Permission is granted to copy, distribute and/or modify this document under the terms of the GNU Free Documentation License, Version 1.2 or any later version published by the Free Software Foundation; with no Invariant Sections, with the Front-Cover Texts, and with the Back- Cover Texts. A copy of the license is included in the section entitled "GNU Free Documentation License".

All parts of this book are extracted from Wikipedia, the free encyclopedia (www.wikipedia.org).

You can get detailed informations about the authors of this collection of articles at the end of this book. The editors (Ed.) of this book are no authors. They have not modified or extended the original texts.

Pictures published in this book can be under different licences than the GNU Free Documentation License. You can get detailed informations about the authors and licences of pictures at the end of this book.

The content of this book was generated collaboratively by volunteers. Please be advised that nothing found here has necessarily been reviewed by people with the expertise required to provide you with complete, accurate or reliable information. Some information in this book maybe misleading or wrong. The Publisher does not guarantee the validity of the information found here. If you need specific advice (f.e. in fields of medical, legal, financial, or risk management questions) please contact a professional who is licensed or knowledgeable in that area.

Any brand names and product names mentioned in this book are subject to trademark, brand or patent protection and are trademarks or registered trademarks of their respective holders. The use of brand names, product names, common names, trade names, product descriptions etc. even without a particular marking in this works is in no way to be construed to mean that such names may be regarded as unrestricted in respect of trademark and brand protection legislation and could thus be used by anyone.

Cover image: www.ingimage.com
Concerning the licence of the cover image please contact ingimage.

Publisher:
Part Press is a trademark of
International Book Market Service Ltd., 17 Rue Meldrum, Beau Bassin, 1713-01 Mauritius
Email: info@bookmarketservice.com
Website: www.bookmarketservice.com

Published in 2013

Printed in: U.S.A., U.K., Germany. This book was not produced in Mauritius.

ISBN: 978-613-8-59130-6

Contents

Articles

References

Rogóźno, Łask County

Rogóźno	
— Village —	

Rogóźno

Coordinates: 51°26′5″N 18°58′42″E

Country	Poland
Voivodeship	Łódź
County	Łask
Gmina	Widawa
Population	420

Rogóźno [rɔˈɡuʑnɔ] is a village in the administrative district of Gmina Widawa, within Łask County, Łódź Voivodeship, in central Poland.[1] It lies approximately 3 kilometres (2 mi) east of Widawa, 20 km (12 mi) south-west of Łask, and 52 km (32 mi) south-west of the regional capital Łódź.

The village has a population of 420.

References

[1] "Central Statistical Office (GUS) – TERYT (National Register of Territorial Land Apportionment Journal)" (http://www.stat.gov.pl/broker/access/prefile/listPreFiles.jspa) (in Polish). 2008-06-01. .

Łódź

Łódź

Top: Freedom Square and Kościuszko monument
Upper Middle: Manufaktura Centre, Poznański Palace
Lower Middle: Piotrkowska Street, Łódź Film School
Bottom: Panorama of Łódź at dusk

Flag

Coat of arms

Motto: *Ex navicula navis* (*From a boat, a ship*)

Łódź

Coordinates: 51°47′N 19°28′E

Country	Poland
Voivodeship	Łódź
County	*city county*
City Rights	1423

Government	
• Mayor	Hanna Zdanowska
Area	
• City	293.25 km^2 (**unknown operator: u'strong'** sq mi)
Highest elevation	278 m (912 ft)
Lowest elevation	162 m (531 ft)
Population (2011)	
• City	728 892
• Metro	1428600
Time zone	CET (UTC+1)
• Summer (DST)	CEST (UTC+2)
Postal code	90-001 to 94–413
Area code(s)	+48 42
Car plates	EL
Website	http://www.uml.lodz.pl/

Łódź (Polish pronunciation: [wutɕ] (listen); Yiddish: לאָדזש, *Lodzh*; English pronunciation: /luːdʒ/ or /lɒdz/[1]) is the third-largest city in Poland. Located in the central part of the country, it had a population of 742,387 in December 2009. It is the capital of Łódź Voivodeship, and is approximately 135 kilometres (**unknown operator: u'strong'** mi) south-west of Warsaw. The city's coat of arms is an example of canting: depicting a boat, it alludes to the city's name which translates literally as "boat".

History

Łódź first appears in the written record in a 1332 document giving the village of *Łodzia* to the bishops of Włocławek. In 1423 King Władysław Jagiełło granted city rights to the village of *Łódź*. From then until the 18th century the town remained a small settlement on a trade route between Masovia and Silesia. In the 16th century the town had fewer than 800 inhabitants, mostly working on the nearby grain farms.

With the second partition of Poland in 1793, Łódź became part of the Kingdom of Prussia's province of South Prussia, and was known in German as *Lodsch*. In 1798 the Prussians nationalised the town, and it lost its status as a town of the bishops of Kuyavia. In 1806 Łódź joined the Napoleonic Duchy of Warsaw and in 1810 it had 190 inhabitants. In the 1815 Congress of Vienna treaty it became part of Congress Poland, a client state of the Russian Empire.

Century of partitions

St Stanisław Kostka Arch-Cathedral Basilica in Łódź, 1901

In the 1815 treaty, it was planned to renew the dilapidated town and with the 1816 decree by the Czar a number of German immigrants received territory deeds for them to clear the land and to build factories and housing. In 1820 Stanisław Staszic aided in changing the small town into a modern industrial centre. The immigrants came to the *Promised Land* (*Ziemia obiecana*, the city's nickname) from all over Europe. Mostly they arrived from Southern Germany, Silesia and Bohemia, but also from countries as far away as Portugal, England, France and Ireland. The first cotton mill opened in 1825, and 14 years later the first steam-powered factory in both Poland and Russia commenced operations. In 1839 the population was 80% Germans and German schools and churches were established.

A constant influx of workers, businessmen and craftsmen from all over Europe transformed Łódź into the main textile production centre of the Russian Empire. Three groups dominated the city's population and contributed the most to the city's development: Poles, Germans and Jews, who started to arrive since 1848. Many of the Łódź craftspeople were weavers from Silesia.

In 1850, Russia abolished the customs barrier between Congress Poland and Russia proper; industry in Łódź could now develop freely with a huge Russian market not far away. Soon the city became the second-largest city of Congress Poland. In 1865 the first railroad line opened (to Koluszki, branch line of the Warsaw-Vienna Railway), and soon the city had rail links with Warsaw and Białystok.

One of the most important industrialists of Łódź was Karl Wilhelm Scheibler.[2] In 1852 he came to Łódź and with Julius Schwarz together started buying property and building several factories. Scheibler later bought out Schwarz's share and thus became sole owner of a large business. After he died in 1881 his widow and other members of the family decided to pay homage to his memory by erecting a chapel, intended as a mausoleum with family crypt, in the Lutheran part of the Łódź cemetery in ulica Ogrodowa (later known as The Old Cemetery).[3]

Historical population

Year	Inhabitants
1793	190
1806	767
1830	4,300
1850	15,800
1880	77,600
1905	343,900
1925	538,600
1990	850,000
2003	781,900
2007	753,192
2009	742,387

In the 1823–1873, the city's population doubled every ten years. The years 1870–1890 marked the period of most intense industrial development in the city's history. Many of the industrialists were Jewish. Łódź soon became a major centre of the socialist movement. In 1892 a huge strike paralysed most of the factories. According to Russian census of 1897, out of the total population of 315,000, Jews constituted 99,000 (around 31% percent).[4] During the 1905 Revolution, in what became known as the June Days or Łódź insurrection, Tsarist police killed more than 300 workers.

Despite the air of impending crisis preceding World War I, the city grew constantly until 1914. By that year it had become one of the most densely-populated industrial cities in the world—13280 inhabitants per square kilometre (**unknown operator: u'strong'** /sq mi). A major battle was fought near the city in late 1914, and as a result the city came under German occupation after 6 December but with Polish independence restored in November 1918 the local population liberated the city and disarmed the German troops. In the aftermath of World War I, Łódź lost approximately 40% of its inhabitants, mostly owing to draft, diseases and because a huge part of the German population was forced to move to Germany.

In sovereign Poland

In 1922, Łódź became the capital of the Łódź Voivodeship, but the period of rapid growth had ceased. The Great Depression of the 1930s and the *Customs War* with Germany closed western markets to Polish textiles while the Bolshevik Revolution (1917) and the Civil War in Russia (1918–1922) put an end to the most profitable trade with the East. The city became a scene of a series of huge workers' protests and riots in the interbellum.

On 13 September 1925 a new airport, Lublinek Airport, began operations on the outskirts of the city. In the interwar years Łódź continued to be a diverse and multicultural city, with the 1931 Polish census showing that the total population of 604,470 included 315,622 (52.21%) Poles, 202,497 (33.49%) Jews and 86,351 (14.28%) Germans (determination based on the declaration of language used) and others.

Occupation by Nazi Germany

During the Invasion of Poland the Polish forces of the Łódź Army of General Juliusz Rómmel defended Łódź against initial German attacks. However, the Wehrmacht captured the city on 8 September. Despite plans for the city to become a Polish exclave, attached to the General Government, the Nazi hierarchy respected the wishes of the local governor of Reichsgau Wartheland, Arthur Greiser, and of many of the ethnic Germans living in the city, and annexed it to the Reich in November 1939. The city received the new name of *Litzmannstadt* after the German general Karl Litzmann, who captured the city during World War I. Nevertheless, many Łódź Germans refused to sign Volksliste and become Volksdeutsche, instead being deported to the General Government.

Memorial to Holocaust victims at Radegast railway station

Soon the Nazi authorities set up the Łódź Ghetto in the city and populated it with more than 200,000 Jews from the Łódź area.[5] As Jews were deported from Litzmannstadt for extermination others were brought in. Due to the value of the goods that the ghetto population produced for the German military and various civilian contractors it was the last major ghetto to be "liquidated" (destroyed); approximately 900 people survived the liquidation of the ghetto in August 1944. Several concentration camps and death camps arose in the city's vicinity for the non-Jewish inhabitants of the regions, among them the infamous Radogoszcz prison and several minor camps for the Romani people and for Polish children.

Thousands of new ethnic-German Volksdeutsche came to Łódź from across Eastern Europe, many of whom were repatriated from Russia during the time of Hitler's alliance with the Soviet Union, and especially after the Operation Barbarossa. In January 1945 most of the German population fled the city for fear of the Red Army. The city also suffered tremendous losses due to the German policy of requisition of all factories and machines and transporting them to Germany. Thus despite relatively small losses due to aerial bombardment and the fighting, Łódź had lost most of its infrastructure.

Prior to World War II, the Jewish population of Łódź numbered about 233,000, accounting for one-third of the city's population. The community was wiped out in the Holocaust. By the end of World War II, Łódź had lost approximately 420,000 of its pre-war inhabitants including 300,000 Polish Jews from Łódź and its vicinity and approximately 120,000 other Poles.[6]

The Soviet Red Army entered the city on 18 January 1945. According to Marshal Katukov, whose forces participated in the operation, the Germans retreated so suddenly that they had no time to evacuate or destroy the Łódź factories, as they did in other cities.[7] In time, Łódź became part of the People's Republic of Poland.

After World War II

At the end of World War II, Łódź had fewer than 300,000 inhabitants. However the number began to grow as refugees from Warsaw and territories annexed by the Soviet Union immigrated. Until 1948 the city served as a *de facto* capital of Poland, since events during and after the Warsaw uprising had thoroughly destroyed Warsaw, and most of the government and country administration resided in Łódź. Some planned moving the capital there permanently, however this idea did not gain popular support and in 1948 the reconstruction of Warsaw began. Under the Polish Communist regime many of the industrialist families lost their wealth when the authorities nationalised private companies. Once again the city became a major centre of industry. In mid-1981 Łódź became famous for its massive, 50,000 hunger demonstration of local mothers and their children (see: Summer 1981 hunger demonstrations in Poland). After the period of economic transition during the 1990s, most enterprises were again privatised.

Fountain on Jan Henryk Dąbrowski Square

Piotrkowska Street

Economy

Before 1990, Łódź's economy focused on the textile industry, which in the nineteenth century had developed in the city owing to the favourable chemical composition of its water. Because of the growth in this industry, the city has sometimes been called the "Polish Manchester". As a result, Łódź grew from a population of 13,000 in 1840 to over 500,000 in 1913. By the time right before World War I Łódź had become one of the most densely populated industrial cities in the world, with 13,280 inhabitants per km^2. The textile industry declined dramatically in 1990 and 1991, and no major textile company survives in Łódź today. However, countless small companies still provide a significant output of textiles, mostly for export to Russia and other countries of the former Soviet Union.

Sienkiewicza Street

The city benefits from its central location in Poland. A number of firms have located their logistics centres in the vicinity. Two motorways, A1 spanning from the north to the south of Poland, and A2 going from the east to the west intersect northeast of the city. As of 2012, A2 is complete towards Warsaw, while the northern section A1 is largely completed. With these connections, the advantages due to the city's central location should increase even further. Work has also begun on upgrading the railway connection with Warsaw, which reduces the 2 hour travel time to make the 137 km (85 mi) journey to 1.5 hours in 2009. In the next few years much of the track will be modified to handle trains moving at 160 km/h (99 mph), cutting the travel time to about 75 minutes.

Recent years has seen many foreign companies opening offices in Łódź. Indian IT company Infosys has one of its centres in Łódź. Despite the fact that Łódź is regarded to be the poorest among Polish cities with population over 500,000, the GDP per capita in Łódź was 123,9% of Poland's average (2008).

In January 2009 Dell announced that it will shift production from its plant in Limerick, Ireland to its plant in Łódź, largely because the labour costs in Poland are a fraction of those in Ireland.[8] The city's investor friendly policies have attracted 980 foreign investors by January 2009.[8] Foreign investment was one of the factors which decreased the unemployment rate in Łódź to 6.5 percent in December 2008, from 20 percent four years earlier.[8]

Tourism

Sculpture of Artur Rubinstein on Piotrkowska Street in Łódź, where Rubinstein was born and raised

Piotrkowska Street is the main artery and attraction stretching north to south for a little over five kilometres, making it (one of) the longest commercial streets in the world. A few of the building fronts have been renovated and date back to the 19th century.

Although Łódź does not have any hills nor any large body of water, one can still get close to nature in one of the city's many parks, most notably Łagiewniki (the largest city park in Europe). Łódź has one of the best museums of modern art in Poland, *Muzeum Sztuki* on Więckowskiego Street, which displays art by all important contemporary Polish artists. Despite insufficient exhibition space (many very impressive paintings and sculptures lie in storage in the basement), there are plans to move the museum to a larger space in the near future. There is also a branch of *Muzeum Sztuki* called *MS*^{2} located in the area of Łódź largest mall "Manufaktura".

Another popular source of recreation is the Lunapark, an amusement park featuring about two dozen attractions including an 18 metre tall roller coaster and two dozen other rides and features, located near the city's zoo and its botanical gardens.

The largest 19th Century textile factory complex which was built by Izrael Poznanski has been turned into a shopping centre called "Manufaktura" which is an example of a modern business which operates in restored nineteenth century buildings.

Education

Łódź is a thriving center of academic life. Currently Łódź hosts three major state-owned universities, six higher education establishments operating for more than a half of the century, and a number of smaller schools of higher education. The tertiary institutes with the most students in Łódź include:

Technical University rector's office (formerly Richter's residence, 1904)

- University of Łódź (*Uniwersytet Łódzki*)
- Technical University of Łódź (*Politechnika Łódzka*)
- Medical University of Łódź (*Uniwersytet Medyczny w Łodzi*)
- National Film School in Łódź (*Państwowa Wyższa Szkoła Filmowa, Telewizyjna i Teatralna w Łodzi*)
- Academy of Music in Łódź (*Akademia Muzyczna im. Grażyna i Kiejstuta Bacewiczów w Łodzi*)
- Academy of Fine Arts and Design (*Akademia Sztuk Pięknych im. Wł. Strzemińskiego w Łodzi*)

There are also dozens of other schools and academies, but for the last four years the best students in Łódź (according to the prestigious contest "Studencki Nobel") have been studying at the University of Łódź - in 2009 the regional laureate was Piotr Pawlikowski, in 2010 - Joanna Dziuba, in 2011 and 2012 - Paweł Rogaliński.[9] [10] In recent years, an estimated number of 15 private higher education establishments were created, which gives students better educational opportunities than before.

National Film School in Łódź

The *Leon Schiller's National Higher School of Film, Television and Theatre* in Łódź (*Państwowa Wyższa Szkoła Filmowa, Telewizyjna i Teatralna im. Leona Schillera w Łodzi*) is the most notable academy for future actors, directors, photographers, camera operators and TV staff in Poland. It was founded on 8 March 1948 and was initially planned to be moved to Warsaw as soon as the city was rebuilt following the Warsaw Uprising. However, in the end the school remained in Łódź and today is one of the best-known institutions of higher education in that town.

National Film School at Oskar Kon Palace

At the end of the Second World War Łódź remained the only large Polish town besides Kraków which war had not destroyed. The creation of the National Film School gave the town a role of greater importance from a cultural viewpoint, which before the war had belonged exclusively to Warsaw and Kraków. Early students of the School include the directors Andrzej Munk, Andrzej Wajda, Kazimierz Karabasz (one of the founders of the so-called Black Series of Polish Documentary) and Janusz Morgenstern, who at the end of the Fifties became famous as one of the founders of the Polish Film School of Cinematography.

Immediately after the war, Jerzy Bossak, Wanda Jakubowska, Stanisław Wohl, Antoni Bohdziewicz and Jerzy Toeplitz worked as the first teachers. The internationally renowned film director Roman Polański was among the many talented students who attended the School in the 1950s. Łódź's cinematic involvement and its Hollywood-style star walk on Piotrkowska Street have earned it the nickname "Holly-Łódź". The school is also associated with the Camerimage Film Festival, which occurs annually in late November and early December. Founded in Toruń in 1993, the festival was specifically organised to focus on the art of cinematography and is well-attended every year by world-renowned cinematographers, many of whom also participate in seminars, workshops, retrospectives and Q&A

sessions. Because of both subject matter and attendee composition, it is considered a key event for industry exhibitors, who often make European debuts of their products here.

Łódź in literature and cinema

Three major novels depict the development of industrial Łódź: Władysław Reymont's *The Promised Land* (1898), Joseph Roth's *Hotel Savoy* (1924) and Israel Joshua Singer's *The Brothers Ashkenazi* (1937). Roth's novel depicts the city on the eve of a workers' riot in 1919. Reymont's novel was made into a film by Andrzej Wajda in 1975: see *The Promised Land*. In the 1990 film *Europa Europa*, Solomon Perel's family flees pre-WWII Berlin and settles in Łódź. Scenes of David Lynch's 2006 film *Inland Empire* were shot in Łódź. Sections of Harry Turtledove's *Worldwar* alternate history series take place in Łódź.

Sports

Atlas Arena in Łódź

- BBRC Łódź[11] – rugby union (youth, men, women)- 3rd place in Polish Rugby7 League in 2009/2010
- Budowlani Łódź- winner of Polish Rugby League in 1982/1983, 2005/2006, 2006/2007 and 2009/2010
- ŁKS Łódź – men's football team (established in 1908), (Polish Champion 1958, 1998; Polish Cup winner 1957; First League
- ŁKS Łódź – women's basketball team, 6th place in Sharp Torell Basket Liga in 2003–2004 season
- KS Społem Łódź – leading youth road and track cycling team in Poland
- Klub Żużlowy Orzeł Łódź – speedway team (established in present form in 2005)
- Torpedy Łódź – American football team made to the Final of the 2nd League play-offs in 2008
- Widzew Łódź – men's football team (established in 1910), (Polish Champion 1981, 1982, 1996, 1997; Polish Cup winner: 1985; Polish SuperCup winner: 1996;T-Mobile Ekstraklasa

Government

Łódź constituency

Members of Parliament (Sejm) elected from Łódź constituency:

- Wojciech Olejniczak, LiD
- Zdzisława Janowska, LiD
- Mirosław Drzewiecki, PO
- Iwona Śledzińska-Katarasińska, PO
- John Abraham Godson, PO
- Cezary Grabarczyk, PO
- Joanna Skrzydlewska, PO
- Piotr Krzywicki, PiS
- Jarosław Jagiełło, PiS

Members of Parliament (Senat) elected from Łódź constituency:

- Krzysztof Kwiatkowski, PO
- Maciej Grubski, PO

Mayor

- Waldemar Bohdanowicz, Solidarity (November 1989 – 1990) – appointed by Prime Minister Tadeusz Mazowiecki.
- Grzegorz Palka (1990–1994)
- Marek Czekalski, Freedom Union (1994–1998)
- Tadeusz Matusiak, SLD (1998–2001)
- Krzysztof Panas, SLD (2001–2002)
- Krzysztof Jagiełło, SLD (2002)
- Jerzy Kropiwnicki, Christian-National Union (ZChN) (2002–2010)
- Tomasz Sadzyński, Platforma Obywatelska / Civic Platform (temporary in 2010)
- Hanna Zdanowska, Platforma Obywatelska / Civic Platform

Consulates

Łódź is home to eight foreign consulates, all of which are Honorary. They are subordinate to the following states main representation in Poland: French, Danish, German, Austrian, British, Belgian, Latvian, and Hungarian.

International relations

Twin towns – sister cities

Łódź is twinned with:[12]

• Chemnitz in Germany *(since 1972)*	• Rustavi in Georgia *(since 1995)*
• Stuttgart in Germany *(since 1988)*[13]	• Barreiro in Portugal *(since 1996)*
• Lyon in France *(since 1991)*[14]	• Tampere in Finland *(since 1996)*
• Vilnius in Lithuania *(since 1991)*	• Puebla in Mexico *(since 1997)*
• Ivanovo in Russia *(since 1992)*	• Murcia in Spain *(since 1999)*
• Kaliningrad in Russia *(since 1992)*[15]	• Örebro in Sweden *(since 2001)*[16]
• Minsk in Belarus *(since 1993)*[17]	• Lviv in Ukraine "(since 2003)"
• Odessa in Ukraine *(since 1993)*	• Szeged in Hungary *(since 2004)*
• Tel-Aviv in Israel *(since 1994)*[18]	• Córdoba in Argentina
• Tianjin in People's Republic of China *(since 1994)*	• Los Angeles in the United States *(since 2010)*[19]

Łódź belongs also to the Eurocities network.

Notable residents

- Daniel Amit, Israeli physicist
- Grażyna Bacewicz, composer
- Aleksander Bardini, stage director and actor
- Andrzej Bartkowiak, cameraman and film director
- Jurek Becker (1937–1997) writer
- Kazimierz Brandys, writer
- Artur Brauner, film producer
- Jacob Bronowski, writer, mathematician, and Britain's leading academic TV figure of the 1970s.
- Sabina Citron, Holocaust survivor, activist, and author
- Karl Dedecius, translator
- Karl Dominik (Born:Karol Dominik Ignaczak), China's first Chinese speaking Polish actor

- Max Factor, Sr., businessman, founder of the Max Factor cosmetics company
- Piotr Fronczewski, Polish actor
- Marcin Gortat, NBA basketball player for the Phoenix Suns
- Mendel Grossman, Łódź ghetto photographer[20]
- Józef Hecht (1891–1951), engraver and printmaker
- Josef Joffe, journalist
- Jan Karski, diplomat and antinazi resistant
- Aharon Katzir (1914–72), Israeli pioneer in study of electrochemistry of biopolymers; killed in Lod Airport Massacre
- Lea Koenig, Israeli actress
- Paul Klecki, conductor
- Katarzyna Kobro, sculptor
- Jerzy Kosinski, writer
- Jan Kowalewski, Polish cryptologist who broke Soviet military codes and ciphers during the Polish-Soviet War
- Feliks W. Kres, fantasy writer
- Daniel Libeskind, architect
- Tadeusz Miciński, poet
- Zew Wawa Morejno, Chief Rabbi
- Zbigniew Nienacki, writer
- Josef Olechowski, Lawyer, Polish Senator, Anti-Soviet counter-espionage operative
- Marian P. Opala, Oklahoma Supreme Court Justice
- Adam Ostrowski, better known as O.S.T.R., rapper
- Władysław Pasikowski, director
- Roman Polanski, cinema director, Oscar and Golden Palm winner
- Zbigniew Rybczyński, animator and Oscar winner
- Władysław Reymont, writer, Nobel Prize winner
- Paweł Rogaliński, journalist
- Joseph Rotblat, Nobel Prize winner
- Stefan Rozental, nuclear physicist
- Arnold Rutkowski, opera singer
- Artur Rubinstein, pianist, settled
- Marek Saganowski, football player
- Andrzej Sapkowski, fantasy writer
- Carl Wilhelm Scheibler (1820–1881) one of the most important Łódź industrialists
- Piotr Sobociński, cinematographer
- Andrzej Sontag, track-and-field star
- Natan Spigel, painter (1900–1942)
- Władysław Strzemiński, painter, Kobro's husband
- Arthur Szyk, artist
- Aleksander Tansman, composer and pianist
- Jack Tramiel, computer manufacturer, founder of Commodore and Atari
- Julian Tuwim, poet
- Miś Uszatek, cartoon character
- Michał Wiśniewski, singer
- Hanna Zdanowska, politician
- Aleksandra Ziółkowska-Boehm, writer

Other buildings

Łódź City Hall, formerly Heinzel Palace

Old Town Square

European Institute, formerly Schweikert Palace

Andel's Hotel, near Manufaktura shopping mall

Music Academy, formerly Poznanski Palace

White Factory

Great Synagogue (Łódź) destroyed by Nazi Germans

A map of *Oskara Flatta* area of Łódź in 1853

Sigillum oppidi Lodzia, 1577

Izrael Poznanski's Palace

Alexander Nevsky Orthodox church

Karol Scheibler's chapel, Old Evangelical–Augsburg church and cemetery

Concert hall

Poniatowski's Park

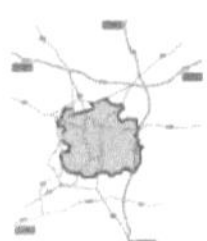

Planned motorway network around Łódź

References

[1] http://www.merriam-webster.com/dictionary/lodz Merriam-Webster online.
[2] "Neues Leben in alten Fabriken: Lódz baut auf Kultur" (http://www.weser-kurier.de/Artikel/Ratgeber/Reise/36989/Neues+Leben+in+alten+Fabriken:+Lodz+baut+auf+Kultur.html) (in (German)). Weser Kurier. 22 September 2009. . Retrieved 2 Oct. 2009.
[3] "Foundation For Saving Karol Scheibler's Chapel" (http://www.scheibler.org.pl/eng_fundation.htm). Scheibler.org.pl. . Retrieved 25 Jan. 2010.
[4] Joshua D. Zimmerman, *Poles, Jews, and the politics of nationality*, Univ of Wisconsin Press, 2004, ISBN 0-299-19464-7, Google Print, p.16 (http://books.google.com/books?id=6sbr9cZyw_4C&pg=PA16&dq=population+Brest+Poles+Jews&lr=&as_brr=3&ei=HS_3SZ2-NITyzQSUtaWtBQ)
[5] Isaiah Trunk: 2006, Page xi
[6] Weiner, Rebecca. The Virtual Jewish History Tour (http://www.jewishvirtuallibrary.org/jsource/vjw/Lodz.html#The Holocaust). American-Israeli Cooperative Enterprise. Retrieved on 15 January 2008.
[7] Blobaum, Robert. " On Strike on Łódź (http://books.google.com/books?vid=ISBN0801432871&id=z_Z_nJnzTp4C&pg=PA75&lpg=PA75&dq=Lodz+1945&sig=TWDJHt_7-aUGw30TT8Eb6C0THiU). "Rewolucja: Russian Poland, 1904–1907". Cornell University Press, 1995. p. 75.
[8] (AFP)–24 Jan 2009 (24 Jan, 2009). "AFP: Dell seeks refuge in Poland as crisis bites" (http://www.google.com/hostednews/afp/article/ALeqM5jfg1UcJtfo3IzRq-Dne3j2H1cSoA). Google.com. . Retrieved 11 Apr, 2010.
[9] http://www.studenckinobel.pl/index.php/historia-konkursu History of the contest "Studencki Nobel" (in Polish)
[10] "Młody dziennikarz znów pretenduje do Nobla! (in Polish)" (http://www.halolodz.pl/wiadomosci/wiadomoÅci/artykul,9369,mÅody-dziennikarz-znÃ³w-pretenduje-do-nobla.html). . Retrieved 2010-05-29.
[11] Oficial website (http://bbrclodz.pl/)
[12] *"Twin Cities"* (http://en.www.uml.lodz.pl/index.php?str=2029). *The City of Łódź Office. © 2007 UMŁ. . Retrieved 23 Oct. 2008.*
[13] Stuttgart official website, 23 Mar 2008 (http://stuttgart.de/sde/menu/frame/ns_top_11021_11041.htm)
[14] "*Partner Cities of Lyon and Greater Lyon*" (http://www.lyon.fr/vdl/sections/en/villes_partenaires/villes_partenaires_2/?aIndex=1). © 2008 Mairie de Lyon. . Retrieved 21 Oct. 2008.
[15] "*Kaliningrad -Partner Cities*" (http://www.klgd.ru/en/search/index.php?q=partner+cities&where=). © 2000–2006 Kaliningrad City Hall. . Retrieved 8 Dec. 2008.
[16] *Vänorter* – orebro.se (http://www.orebro.se/2444.html)
[17] "*Twin towns of Minsk*" (http://minsk.gov.by/cgi-bin/org_ps.pl?k_org=3604&mode=doc&doc=3604_2_a&lang=eng). © 2008 The department of protocol and international relations of Minsk City Executive Committee. . Retrieved 8 Dec. 2008.
[18] "Tel Aviv sister cities" (http://www.tel-aviv.gov.il/Hebrew/Cityhall/TwinCities/Index.asp) (in Hebrew). Tel Aviv-Yafo Municipality. . Retrieved 19 Jan. 2008.
[19] Sister Cities of Los Angeles (http://sistercitiesofla.com/page1/page60/page60.html)
[20] *Holocaust chronicles... – Google Books* (http://books.google.com/?id=dnMMxHaUzT8C&pg=PA125&lpg=PA125&dq=mendel+grossman+photos). Books.google.com. 1999-05. ISBN 978-0-88125-630-7. . Retrieved 25 Jan. 2010.

- Horwitz, Gordon J. (2008). *Ghettostadt: Łódź and the making of a Nazi city* (http://books.google.com/?id=JL-ujgqHlEcC). Harvard University Press, Cambridge, Massachusetts. ISBN 978-0-674-02799-2. Retrieved 6 March 2010.
- Stefański, Krzysztof (2000). *Gmachy użyteczności publicznej dawnej Łodzi*, Łódź 2000 ISBN 83-86699-45-0.
- Stefański, Krzysztof (2009). *Ludzie którzy zbudowali Łódź Leksykon architektów i budowniczych miasta (do 1939 roku)*, Łódź 2009 ISBN 978-83-61253-44-0.
- Trunk, Isaiah; Shapiro, Robert Moses (2006). *Łódź Ghetto: a history* (http://books.google.com/?id=ugVsNrbMSx4C). Indiana University Press, Bloomington, Indiana. ISBN 978-0-253-34755-8. Retrieved 6 March 2010.

Bibliography

- *Ghettostadt: Łódź and the Making of a Nazi City,* Gordon H Horwitz, Harvard University Press, 2009
- "A Stairwell in Lodz," Constance Cappel, 2004, Xlibris, (in English).
- "Lodz – The Last Ghetto in Poland," Michal Unger, Yad Vashem, 600 pages (in Hebrew)

External links

- Official website (http://http://www.lodz.pl/)
- Public Transport Official Site (http://www.mpk.lodz.pl/)
- City map of Łódź (http://www.mapa.lodz.pl/eng/)
- Historic images of Łódź (http://www.bfcollection.net/cities/poland/lodz/lodz_01.html)
- Photo Gallery of Łódź (http://lodz.pdg.pl/)
- Łódź Special Economic Zone (http://www.sse.lodz.pl/)
- Łódź-Lublinek Airport (http://www.airport.lodz.pl/)
- The Łódź (http://www.thelodz.com/) English language newspaper
- The Łódź Ghetto (http://www.holocaustresearchproject.net/ghettos/Lodz/lodzghetto.html)
- Promo website of Lodz (RU) (http://www.lodz-ru.com/)
- Łódź 2016 – Official site of candidate for European Capital of Culture (http://www.lodz2016.com/)
- – Unofficial site of links to site with inform about Lodz (http://lodz.site90.net/)
- Youth Groups in the Łódź Ghetto (http://www1.yadvashem.org/exhibitions/lodz/1_home.html?WT.mc_id=wiki) – Online Exhibition from Yad Vashem

Warsaw

Warsaw ***Warszawa***
Capital City of Warsaw ***Miasto Stołeczne Warszawa***

Left to right: Financial centre, Royal Castle, Old Town Market Place, Presidential Palace, Wilanów Palace.

Flag

Coat of arms

Motto: *Semper invicta* (Latin "Ever Invincible")

Warsaw

Coordinates: 52°13′56.28″N 21°00′30.36″E

Country	Poland
Voivodeship	Masovian
County	*city county*
City rights	turn of the 12th to 13th century
Boroughs	
Government	
• President	Hanna Gronkiewicz-Waltz (PO)
Area	
• City	517.24 km^2 (**unknown operator: u'strong'** sq mi)
• Metro	4222.79 km^2 (**unknown operator: u'strong'** sq mi)
Elevation	78–116 m (328 ft)
Population (31 December 2011)	
• City	1,708,491[1]
• Density	3304/km^2 (**unknown operator: u'strong'**/sq mi)
• Metro	2,666,278
• Metro density	631.4/km^2 (**unknown operator: u'strong'**/sq mi)
Demonym	Varsovian
Time zone	CET (UTC+1)
• Summer (DST)	CEST (UTC+2)
Postal code	00-001 to 04–999
Area code(s)	+48 22

Car plates	WA, WB, WD, WE, WF, WH, WI, WJ, WK, WN, WT, WU, WW, WX, WY
Website	warszawa.pl [2]

Warsaw (Polish: *Warszawa* [varˈʂava] (listen) (also [varˈʃava]); see also other names) is the capital and largest city of Poland. It is located on the Vistula River, roughly 260 kilometres (**unknown operator: u'strong'** mi) from the Baltic Sea and 300 kilometres (**unknown operator: u'strong'** mi) from the Carpathian Mountains. Its population is estimated at 1,708,491 residents within a greater metropolitan area of 2,666,278 residents, making Warsaw the 9th most populous city proper in the European Union.[1] [3] [4] The area of the city covers 516.9 square kilometres (**unknown operator: u'strong'** sq mi), while the city's agglomeration covers 6100.43 square kilometres (**unknown operator: u'strong'** sq mi).[5]

Warsaw is an Alpha- global city, a major international tourist destination and an important economic hub in Central Europe.[6] [7] [8] It is also known as the "phoenix city" because it has survived many wars throughout its history. Most notably, the city had to be painstakingly rebuilt after the extensive damage it suffered from World War II, during which 85% of its buildings were destroyed.[9] [10] On 9 November 1940 the city was awarded Poland's highest military decoration for heroism, the Virtuti Militari, for the Siege of Warsaw (1939).[11] [12]

Warsaw is the source for naming entities such as Warsaw Confederation, the Warsaw Pact, the Duchy of Warsaw, the Warsaw Convention, the Treaty of Warsaw, the Warsaw Uprising, the Warsaw Ghetto, and the Warsaw Ghetto Uprising. The Warszawianka is widely considered the unofficial anthem of the city.[13]

Etymology and names

Warsaw's name in the Polish language, *Warszawa* (also formerly spelled *Warszewa* and *Warszowa*), means "belonging to Warsz", *Warsz* being a shortened form of the masculine name of Slavic origin Warcisław; see also etymology of Wrocław.[14] Folk etymology attributes the city name to a fisherman Wars and his wife Sawa. According to legend, Sawa was a mermaid living in the Vistula River who Wars fell in love with.[15] Actually, Warsz was a 12th/13th century nobleman who owned a village located at the site of today's Mariensztat neighbourhood.[16] The official city name in full is *miasto stołeczne Warszawa* (English: "The Capital City of Warsaw").[17] A native or resident of Warsaw is known as a *Varsovian*.

Other names for Warsaw include *Warschau* (German and Dutch), װאַרשע/*Varshe* (Yiddish), *Варшава/Varshava* (Russian, Bulgarian, Belorussian, Ukranian), *Varšava* (Slovak, Czech, Latvian, Slovenian, Serbo-Croatian), *Varšuva* (Lithuanian), *Varsó* (Hungarian), *Varsovia* (Spanish and Latin), *Varsovie* (French), *Varsavia* (Italian), *Varsóvia* (Portuguese), ורשה/*Varsha* (Hebrew), *Huáshā* (Mandarin Chinese), and وارسو/فرسوفيا *Wārsō/Farsōfyā* (Arabic).

History

Early history

1411 St. Mary's Church in Warsaw New Town

The first fortified settlements on the site of today's Warsaw were Bródno (9th/10th century) and Jazdów (12th/13th century).[19] After Jazdów was raided, a new similar settlement was established on the site of a small fishing village called Warszowa. The Płock prince Bolesław II of Masovia, established this settlement, the modern Warsaw, about 1300. In the beginning of the 14th century it became one of the seats of the Dukes of Masovia, becoming the capital of Masovia in 1413.[19] 14th-century Warsaw's economy rested on crafts and trade. Upon the extinction of the local ducal line, the duchy was reincorporated into the Polish Crown in 1526.[19]

16th to 18th centuries

In 1529 Warsaw for the first time became the seat of the General Sejm, permanent from 1569.[19] In 1573 the city gave its name to the Warsaw Confederation, formally establishing religious freedom in the Polish–Lithuanian Commonwealth. Due to its central location between the Commonwealth's capitals of Kraków and Vilnius, Warsaw became the capital of the Commonwealth, and of the Polish Crown, in 1596, when King Sigismund III Vasa moved the court from Kraków to Warsaw.[19]

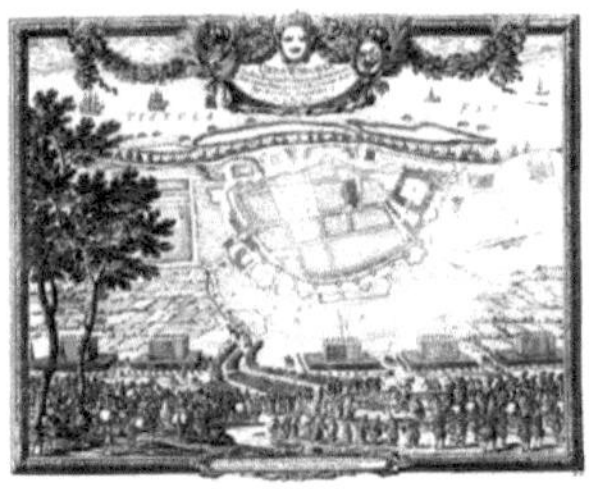

Battle of Warsaw in 1656

In the following years the town expanded towards the suburbs. Several private independent districts were established, the property of aristocrats and the gentry, which were ruled by their own laws. Three times between 1655–1658 the city was under siege and three times it was taken and pillaged by the Swedish, Brandenburgian and Transylvanian forces.[19] [20]

In 1700, the Great Northern War broke out. The city was besieged several times and was obliged to pay heavy contributions.[18] Warsaw turned into an early-capitalistic principal city.

Stanisław August Poniatowski, who remodelled the interior of the Royal Castle in Warsaw, also made Warsaw a centre of culture and the arts.[21] [22] This earned Warsaw the name of *the Paris of the east*.[23]

19th and 20th centuries

Warsaw remained the capital of the Polish–Lithuanian Commonwealth until 1795, when it was annexed by the Kingdom of Prussia to become the capital of the province of South Prussia. Liberated by Napoleon's army in 1806, Warsaw was made the capital of the newly created Duchy of Warsaw.[19] Following the Congress of Vienna of 1815, Warsaw became the centre of the Congress Poland, a constitutional monarchy under a personal union with Imperial Russia.[19] The Royal University of Warsaw was established in 1816.

German airship Schütte Lanz SL2 bombing Warsaw in 1914

Following the repeated violations of the Polish constitution by the Russians, the 1830 November Uprising broke out. However, the Polish-Russian war of 1831 ended in the uprising's defeat and in the curtailment of the Kingdom's autonomy.[19] On 27 February 1861 a Warsaw crowd protesting against the Russian rule over Poland was fired upon by the Russian troops.[24] [25] Five people were killed. The Underground Polish National Government resided in Warsaw during January Uprising in 1863–64.[25]

Warsaw flourished in the late 19th century under Mayor Sokrates Starynkiewicz (1875–92), a Russian-born general appointed by Tsar Alexander III. Under Starynkiewicz Warsaw saw its first water and sewer systems designed and built by the English engineer William Lindley and his son, William Heerlein Lindley, as well as the expansion and modernisation of trams, street lighting and gas works.[19]

> *The history of contemporary civilisation knows no event of greater importance than the Battle of Warsaw, 1920, and none of which the significance is less appreciated.*
>
> —Sir Edgar Vincent d'Abernon[26]

The Russian Empire Census of 1897 recorded 626,000 people living in Warsaw, making it the third-largest city of the Empire after St. Petersburg and Moscow.

Warsaw was occupied by Germany from the 4 August 1915 until 1918. It then became the capital of the newly independent Poland in 1918. In the course of the Polish-Bolshevik War of 1920, the huge Battle of Warsaw was fought on the eastern outskirts of the city in which the capital was successfully defended and the Red Army defeated.[27] Poland stopped by itself the full brunt of the Red Army and defeated an idea of the "export of the revolution".[28]

World War II

Sea of rubble[29] – over eight out of every ten buildings in Warsaw were destroyed by the end of World War II. In left centre can be seen ruins of Old Town Market Square.

During World War II, central Poland, including Warsaw, came under the rule of the General Government, a German Nazi colonial administration. All higher education institutions were immediately closed and Warsaw's entire Jewish population – several hundred thousand, some 30% of the city – herded into the Warsaw Ghetto.[30] The city would become the center of urban resistance to Nazi rule in occupied Europe.[31] When the order came to annihilate the ghetto as part of Hitler's "Final Solution" on 19 April 1943, Jewish fighters launched the Warsaw Ghetto Uprising.[32] Despite being heavily outgunned and outnumbered, the Ghetto held out for almost a month.[32] When the fighting ended, almost all survivors were massacred, with only a few managing to escape or hide.[32] [33]

By July 1944, the Red Army was deep into Polish territory and pursuing the Germans toward Warsaw.[34] Knowing that Stalin was hostile to the idea of an independent Poland, the Polish government-in-exile in London gave orders to the underground Home Army (AK) to try to seize control of Warsaw from the Germans before the Red Army arrived.

Thus, on 1 August 1944, as the Red Army was nearing the city, the Warsaw Uprising began.[34] The armed struggle, planned to last 48 hours, went on for 63 days. Stalin gave orders to his troops to wait outside of Warsaw.[35]

Eventually the Home Army fighters and civilians assisting them were forced to capitulate.[34] They were transported to PoW camps in Germany, while the entire civilian population was expelled.[34] Polish civilian deaths are estimated at between 150,000 and 200,000.[36]

The Germans then razed Warsaw to the ground. Hitler, ignoring the agreed terms of the capitulation, ordered the entire city to be razed to the ground and the library and museum collections taken to Germany or burned.[34] Monuments and government buildings were blown up by special German troops known as *Verbrennungs- und Vernichtungskommando* ("Burning and Destruction Detachments").[34] About 85% of the city had been destroyed, including the historic Old Town and the Royal Castle.[37]

On 17 January 1945 – after the beginning of the Vistula–Oder Offensive of the Red Army – Soviet troops entered the ruins of Warsaw, and liberated Warsaw's suburbs from German occupation. The city was swiftly taken by the Soviet Army, which rapidly advanced towards Łódź, as German forces regrouped at a more westward position.

Modern times

In 1945, after the bombing, the revolts, the fighting, and the demolition had ended, most of Warsaw lay in ruins.

John Paul II's Mass in Victory Square, 1979

After the war, under a Communist regime set up by the conquering Soviets, large prefabricated housing projects were erected in Warsaw to address the housing shortage, along with other typical buildings of an Eastern Bloc city, such as the Palace of Culture and Science. The city resumed its role as the capital of Poland and the country's centre of political and economic life. Many of the historic streets, buildings, and churches were restored to their original form. In 1980, Warsaw's historic Old Town was inscribed onto UNESCO's World Heritage list.[38]

John Paul II's visits to his native country in 1979 and 1983 brought support to the budding solidarity movement and encouraged the growing anti-communist fervor there.[39] In 1979, less than a year after becoming pope, John Paul celebrated Mass in Victory Square in Warsaw and ended his sermon with a call to "renew the face" of Poland: *Let Thy Spirit descend! Let Thy Spirit descend and renew the face of the land! This land!*[39] These words were very meaningful for the Polish citizens who understood them as the incentive for the democratic changes.[39]

In 1995, the Warsaw Metro opened. With the entry of Poland into the European Union in 2004, Warsaw is currently experiencing the biggest economic boom of its history.[40] The opening fixture of UEFA Euro 2012 took place in Warsaw, a game in which the co-hosts Poland, drew 1-1 with Greece.[41]

Geography

Location and topography

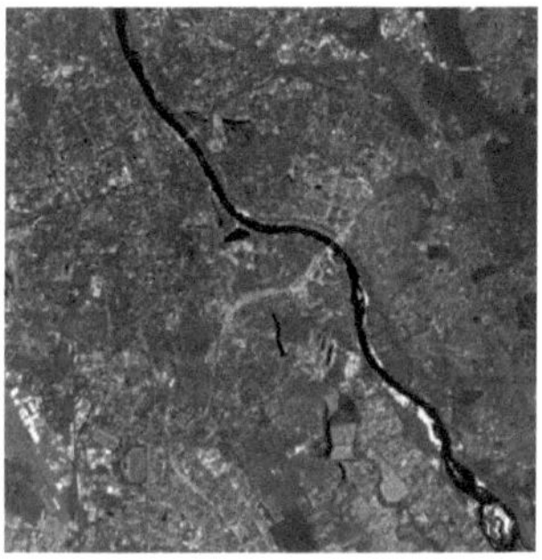

Warsaw seen from SPOT satellite

Warsaw lies in east-central Poland about 300 km (**unknown operator: u'strong'** mi) from the Carpathian Mountains and about 260 km (**unknown operator: u'strong'** mi) from the Baltic Sea, 523 km (**unknown operator: u'strong'** mi) east of Berlin, Germany.[42] The city straddles the Vistula River. It is located in the heartland of the Masovian Plain, and its average elevation is 100 metres (330 ft) above sea level. The highest point on the left side of the city lies at a height of 115.7 metres (379.6 ft) ("Redutowa" bus depot, district of Wola), on the right side – 122.1 metres (400.6 ft) ("Groszówka" estate, district of Wesoła, by the eastern border). The lowest point lies at a height 75.6 metres (248.0 ft) (at the right bank of the Vistula, by the eastern border of Warsaw). There are some hills (mostly artificial) located within the confines of the city – e.g. Warsaw Uprising Hill (121 metres (397.0 ft)), Szczęśliwice hill (138 metres (452.8 ft) – the highest point of Warsaw in general).

Warsaw is located on two main geomorphologic forms: the plain moraine plateau and the Vistula Valley with its asymmetrical pattern of different terraces. The Vistula River is the specific axis of Warsaw, which divides the city into two parts, left and right. The left one is situated both on the moraine plateau (10 to 25 m (32.81 to 82.02 ft) above Vistula level) and on the Vistula terraces (max. 6.5 m (21.33 ft) above Vistula level). The significant element of the relief, in this part of Warsaw, is the edge of moraine plateau called Warsaw Escarpment. It is 20 to 25 m (65.62 to 82.02 ft) high in the Old Town and Central district and about 10 m (32.81 ft) in the north and south of Warsaw. It goes through the city and plays an important role as a landmark.

The plain moraine plateau has only a few natural and artificial ponds and also groups of clay pits. The pattern of the Vistula terraces is asymmetrical. The left side consist mainly of two levels: the highest one contains former flooded terraces and the lowest one the flood plain terrace. The contemporary flooded terrace still has visible valleys and ground depressions with water systems coming from the Vistula old – riverbed. They consist of still quite natural streams and lakes as well as the pattern of drainage ditches. The right side of Warsaw has a different pattern of geomorfological forms. There are several levels of the plain Vistula terraces (flooded as well as former flooded once) and only small part and not so visible moraine escarpment. Aeolian sand with a number of dunes parted by peat swamps or small ponds cover the highest terrace. These are mainly forested areas (pine forest).

Climate

Warsaw's climate is humid continental (Koppen *Dfb*) with cold winters and warm summers, on the border with an oceanic Cfb climate. The average temperature is –3 °C (27 °F) in January and 19.3 °C (66.7 °F) in July. Temperatures may often reach 30 °C (86 °F) in the summer. Yearly rainfall averages 495 millimetres (**unknown operator: u'strong'** in), wettest month being July. Spring and Autumn are usually beautiful seasons, the former crisp and sunny and full of blooms and the latter alternately sunny and misty, and cool but not cold.

Districts

Until 1994, there were 7 districts in Warsaw: Śródmieście, Praga Północ, Praga Południe, Żoliborz, Wola, Ochota, Mokotów. Between 1994 and 2002, there were 11 districts: Centrum, Białołęka, Targówek, Rembertów, Wawer, Wilanów, Ursynów, Włochy, Ursus, Bemowo, Bielany. In 2002, the town Wesoła was incorporated and the territorial division of Warsaw was established as follows:

District	Population	Area
Mokotów	220,682	35.4 km^2 (**unknown operator: u'strong'** sq mi)
Praga Południe	178,665	22.4 km^2 (**unknown operator: u'strong'** sq mi)
Ursynów	145,938	48.6 km^2 (**unknown operator: u'strong'** sq mi)
Wola	137,519	19.26 km^2 (**unknown operator: u'strong'** sq mi)
Bielany	132,683	32.3 km^2 (**unknown operator: u'strong'** sq mi)
Targówek	123,278	24.37 km^2 (**unknown operator: u'strong'** sq mi)
Śródmieście	122,646	15.57 km^2 (**unknown operator: u'strong'** sq mi)
Bemowo	115,873	24.95 km^2 (**unknown operator: u'strong'** sq mi)
Białołęka	96,588	73.04 km^2 (**unknown operator: u'strong'** sq mi)
Ochota	84,990	9.7 km^2 (**unknown operator: u'strong'** sq mi)
Wawer	69,896	79.71 km^2 (**unknown operator: u'strong'** sq mi)
Praga Północ	69,510	11.4 km^2 (**unknown operator: u'strong'** sq mi)
Ursus	53,755	9.35 km^2 (**unknown operator: u'strong'** sq mi)
Żoliborz	48,342	8.5 km^2 (**unknown operator: u'strong'** sq mi)
Włochy	38,075	28.63 km^2 (**unknown operator: u'strong'** sq mi)
Wilanów	23,960	36.73 km^2 (**unknown operator: u'strong'** sq mi)
Rembertów	23,280	19.30 km^2 (**unknown operator: u'strong'** sq mi)
Wesoła	22,811	22.6 km^2 (**unknown operator: u'strong'** sq mi)
Total	1,708,491[1]	521.81 km^2 (**unknown operator: u'strong'** sq mi)

Warsaw is a *powiat* (*county*), and is further divided into 18 boroughs, each one known as a *dzielnica* (*districts* – see map [43]), each one with its own administrative body.[44] Each of the boroughs includes several neighbourhoods which have no legal or administrative status. Warsaw has two historic districts, called Old Town (*Stare Miasto*) and New Town (*Nowe Miasto*) in the borough of Śródmieście.

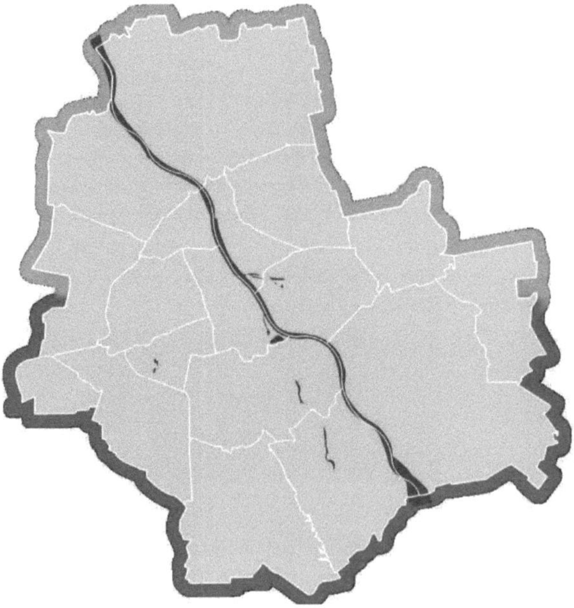

Białołęka

Bielany

Bemowo

Żoliborz

Praga Północ

Targówek

Śródmieście

Wola

Ochota

Włochy

Ursus

Mokotów

Wawer

Praga Południe

Rembertów

Wesoła

Ursynów

Wilanów

Cityscape

Overview

Warsaw's mixture of architectural styles reflects the turbulent history of the city and country. During WWII, Warsaw was razed to the ground by bombing raids and planned destruction.[34] After liberation, rebuilding began as in other cities of the communist-ruled PRL. Most of the historical buildings were thoroughly reconstructed. However, some of the buildings from the 19th century that had been preserved in reasonably reconstructible form were nonetheless eradicated in the 1950s and 1960s (e.g. Leopold Kronenberg Palace).[45] Mass residential blocks were erected, with basic design typical of Eastern bloc countries.

Public spaces attract heavy investment, so that the city has gained entirely new squares, parks and monuments. Warsaw's current urban landscape is one of modern and contemporary architecture.

Architecture

Warsaw's palaces, churches and mansions display a richness of color and architectural details. Buildings are representatives of nearly every European architectural style and historical period. The city has wonderful examples of architecture from the gothic, renaissance, baroque and neoclassical periods, all of which are located within easy walking distance of the town centre.

Łazienki Palace, also called the *Palace on the Water*

Gothic architecture is represented in the majestic churches but also at the burgher houses and fortifications. The most significant buildings are St. John's Cathedral (14th century), the temple is a typical example of the so-called Masovian gothic style, St. Mary's Church (1411), a town house of Burbach family (14th century),[46] Gunpowder Tower (after 1379) and the Royal Castle *Curia Maior* (1407–1410). The most notable examples of Renaissance architecture in the city are the house of Baryczko merchant family (1562), building called "The Negro" (early 17th century) and Salwator tenement (1632). The most interesting examples of mannerist architecture are the Royal Castle (1596–1619) and the Jesuit Church (1609–1626) at Old Town. Among the first structures of the early baroque the most important are St. Hyacinth's Church (1603–1639) and Zygmunt's Column (1644).

Building activity occurred in numerous noble palaces and churches during the later decades of the 17th century. One of the best examples of this architecture are Krasiński Palace (1677–1683), Wilanów Palace (1677–1696) and St. Kazimierz Church (1688–1692). The most impressive examples of rococo architecture are Czapski Palace (1712–1721), Palace of the Four Winds (1730s) and Visitationist Church (façade 1728–1761). The neoclassical architecture in Warsaw can be described by the simplicity of the geometrical forms teamed with a great inspiration from the Roman period. Some of the best examples of the neoclassical style are the Palace on the Water (rebuilt 1775–1795), Królikarnia (1782–1786), Carmelite Church (façade 1761–1783) and Evangelical Holy Trinity Church (1777–1782). The economic growth during the first years of Congress Poland caused a rapid rise architecture. The Neoclassical revival affected all aspects of architecture, the most notable are the Great Theater (1825–1833) and buildings located at Bank Square (1825–1828).

Warsaw University of Technology building courtyard

Bas-relief of an iron forger at MDM neighbourhood, one of prime examples of socialist realism in Polish architecture.

Exceptional examples of the bourgeois architecture of the later periods were not restored by the communist authorities after the war (like mentioned Kronenberg Palace and Insurance Company Rosja building) or they were rebuilt in socialist realism style (like Warsaw Philharmony edifice originally inspired by Palais Garnier in Paris). Despite that the Warsaw University of Technology building (1899–1902)[47] is the most interesting of the late 19th century architecture. Lot of the 19th century buildings is restored in Praga (Vistula's right bank), though they are in a pretty bad condition. Warsaw's municipal government authorities have decided to rebuild the Saxon Palace and the Brühl Palace, the most distinctive buildings in prewar Warsaw.[48]

Notable examples of contemporary architecture include the Palace of Culture and Science (1952–1955), a Soc-realist skyscraper located in the city centre, and the Constitution Square with its monumental Socialist realism architecture (MDM estate).[49] The central part of the right-bank (east) Praga borough it is a place where very run-down houses stand right next to modern apartment buildings and shopping malls.

Modern architecture in Warsaw is represented by the Metropolitan Office Building at Pilsudski Square by Lord Foster,[50] Warsaw University Library (BUW) by Marek Budzyński and Zbigniew Badowski, featuring a garden on its roof and view of the Vistula River, Rondo 1 office building by Skidmore, Owings and Merrill and Golden Terraces, consisting of seven overlapping domes retail and business centre.

It has been said that Warsaw, together with Frankfurt, London, Paris, Moscow, Istanbul and Rotterdam is one of the tallest cities in Europe.[51] Of the 21 tallest skyscrapers in Poland, 18 are situated in Warsaw.

Flora and fauna

Greenspace covers 25% of the surface area of Warsaw,[52] including a broad range of greenstructures, from small neighborhood parks, green spaces along streets and in courtyards, trees and avenues to large historic parks, nature conservation areas and the urban forests at the fringe of the city.

Peacocks – one of a number of animal species in the Royal Baths Park.

There are as many as 82 parks in the city which cover 8% of its area.[53] The oldest ones, once parts of representative palaces, are Saxon Garden, the Krasiński Palace Garden, the Royal Baths Park, the Wilanów Palace Park and the Królikarnia Palace Park (*See also: Greenery in the city*).

The Saxon Garden, covering the area of 15.5 ha, was formally a royal garden. There are over 100 different species of trees and the avenues are a place to sit and relax. At the east end of the park, the Tomb of the Unknown Soldier is situated. In the 19th century the Krasiński Palace Garden was remodelled by Franciszek Szanior. Within the central area of the park one can still find old trees dating from that period: maidenhair tree, black walnut, Turkish hazel and Caucasian wingnut trees. With its benches, flower carpets, a pond with ducks on and a playground for kids, the Krasiński Palace Garden is a popular strolling destination for the Varsovians. The Monument of the Warsaw Ghetto Uprising is also situated here. The Royal Baths Park covers the area of 76 ha. The unique character and history of the park is reflected in its landscape architecture (pavilions, sculptures, bridges, cascades, ponds) and vegetation (domestic and foreign species of trees and bushes). What makes this park different from other green spaces in Warsaw is the presence of peacocks and pheasants, which can be seen here walking around freely, and royal carps in the pond. The Wilanów Palace Park, dates back to the second half of the 17th century. It covers the area of 43 ha. Its central French-styled area corresponds to the ancient, baroque forms of the palace. The eastern section of the park, closest to the Palace, is the two-level garden with a terrace facing the pond. The park around the Królikarnia Palace is situated on the old escarpment of the Vistula. The park has lanes running on a few levels deep into the ravines on both sides of the palace.

19th century New Orangery houses a palm house.

Other green spaces in the city include the Botanic Garden and the University Library garden. They have extensive botanical collection of rare domestic and foreign plants, while a palm house in the New Orangery displays plants of subtropics from all over the world.[54] Besides, within the city borders, there are also: *Pole Mokotowskie* (a big park in the northern Mokotów, where was the first horse racetrack and then the airport), *Park Ujazdowski* (close to the Sejm and John Lennon street), Park of Culture and Rest in Powsin, by the southern city border, *Park Skaryszewski* by the right Vistula bank, in Praga. The oldest park in Praga, the Praga Park, was established in 1865–1871 and designed by Jan Dobrowolski.[55] In 1927 a zoological garden (*Ogród Zoologiczny*) was established on the park grounds,[56] and in 1952 a bear run, still open today.

The flora of the city may be considered very rich in species. The species richness is mainly due to the location of Warsaw within the border region of several big floral regions comprising substantial proportions of close-to-wilderness areas (natural forests, wetlands along the Vistula) as well as arable land, meadows and forests. Bielany Forest, located within the borders of Warsaw, is the remaining part of the Masovian Primeval Forest. Bielany Forest nature reserve is connected with Kampinos Forest.[57] It is home to rich fauna and flora. Within the forest there are three cycling and walking trails. Other big forest area is Kabaty Forest by the southern city border. Warsaw has also two botanic gardens: by the Łazienki park (a didactic-research unit of the University of Warsaw) as well as by the Park of Culture and Rest in Powsin (a unit of the Polish Academy of Science).

There is 13 nature reserves in Warsaw – among others, Bielany Forest, Kabaty Woods, Czerniaków Lake. About 15 km from Warsaw, the Vistula river's environment changes strikingly and features a perfectly preserved ecosystem, with a habitat of animals that includes the otter, beaver and hundreds of bird species.[58] There is also several lakes in Warsaw – mainly the oxbow lakes, like Czerniaków Lake, the lakes in the Łazienki or Wilanów Parks, Kamionek Lake. There are lot of small lakes in the parks, but only part of them is permanent – the most of them is being emptied before winter to clean them of plants and sediments.

The Warsaw Zoo covers an area of 40 hectares (**unknown operator: u'strong'** acres).[59] There are about 5,000 animals representing nearly 500 species.[59] Although officially created in 1928,[59] it traces back its roots to 17th century private menageries, often open to the public.[60] [61]

Demographics

Historical population		
Year	**Pop.**	**±%**
1700	30000	—
1792	120000	+300.0%
1800	63400	−47.2%
1830	139700	+120.3%
1850	163600	+17.1%
1882	383000	+134.1%
1901	711988	+85.9%
1909	764054	+7.3%
1925	1003000	+31.3%
1933	1178914	+17.5%
1939	1300000	+10.3%
1945	422000	−67.5%
1950	803800	+90.5%
1960	1136000	+41.3%
1970	1315600	+15.8%
1980	1596100	+21.3%
1990	1655700	+3.7%
2000	1672400	+1.0%

2002	1688200	+0.9%
2006	1702100	+0.8%
2009	1714466	+0.7%
2010	1720398	+0.3%
2011	1708491	−0.7%
Note: 2006[62] 2010[63] 2011[1]		

Historically, Warsaw has been a destination for internal and foreign immigration, especially from Central and Eastern Europe. For nearly 300 years it was known as the "Paris of the North" or "Second Paris".[29] It was always a centre of European culture, existed as a major European city, and was a destination for many Europeans. Demographically it was the most diverse city in Poland, with a significant numbers of foreign-born inhabitants. In addition to the Polish majority, there was a significant Jewish minority in Warsaw. According to Russian census of 1897, out of the total population of 638,000, Jews constituted 219,000 (around 34% percent).[64] Warsaw's prewar Jewish population of more than 350,000 constituted about 30 percent of the city's total population.[65] In 1933 out of 1,178,914 inhabitants 833,500 were of Polish mother tongue.[66] World War II changed all of this, and to this day there is much less ethnic diversity than in the previous 300 years of the city's history.[65] Most of the modern day population growth is based on internal migration and urbanisation.

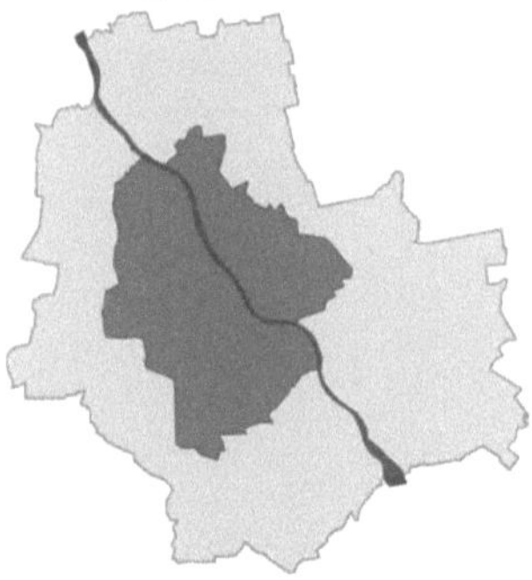

Comparison of Warsaw's city boundaries today and in 1939

In 1939, ca. 1,300,000 people lived in Warsaw,[67] but in 1945 – only 420,000. During the first years after the war, the population growth was ca. 6%, so shortly the city started to suffer from the lack of flats and of areas for new houses. The first remedial measure was the Warsaw area enlargement (1951) – but the city authorities were still forced to introduce residency registration limitations: only the spouses and children of the permanent residents as well as some persons of public importance (like renowned specialists) were allowed to get the registration, hence halving the population growth in the following years. It also bolstered some kind of conviction among Poles that Varsovians thought of themselves as better only because they lived in the capital. Unfortunately this belief still lives on in Poland (although not as much as it used to be) – even though since 1990 there are no limitations to residency registration anymore .[68] [69]

Municipal government

The municipal government existed in Warsaw until World War II and was restored in 1990 (during the communist times, the National City Council – *Miejska Rada Narodowa* – governed in Warsaw). Since 1990, the system of city administration has been changed several times – also as the result of the reform which restored *powiats*, cancelled in 1975. Finally, according the *Warsaw Act*, the city is divided into 18 districts and forms one *city powiat* with a unified municipal government.[70]

Commission Palace, a neoclassical palace at the Bank Square, the house of the city's government.

The basic unit of territorial division in Poland is a commune (*gmina*).[71] A city is also a commune – but with the city charter.[71] Both cities and communes are being governed by a mayor – but in the communes the mayor is vogt (*wójt* in Polish), however in the cities – *burmistrz*. Some bigger cities obtain the entitlements, i.e. tasks and privileges, which are possessed by the units of the second level of the territorial division – counties or *powiat*s. An example of such entitlement is a car registration: a *gmina* cannot register cars, this is a *powiat's* task (i.e. a registration number depends on what *powiat* a car had been registered, not *gmina*). In this case we say about city county or *powiat grodzki*. Such cities are for example Kraków, Gdańsk, Poznań. In Warsaw, its districts additionally have some of *powiat's* entitlements – like already mentioned car registration. For example, the district Wola has its own evidence and the district Ursynów – its own (and the cars from Wola have another type of registration number than these from Ursynów). But for instance the districts in Kraków do not have entitlements of *powiat*, so the registration numbers in Kraków are of the same type for all districts.

Legislative power in Warsaw is vested in a unicameral Warsaw City Council (*Rada Miasta*), which comprises 60 members.[70] Council members are elected directly every four years. Like most legislative bodies, the City Council divides itself into committees which have the oversight of various functions of the city government.[70] Bills passed by a simple majority are sent to the mayor (the President of Warsaw), who may sign them into law. If the mayor vetoes a bill, the Council has 30 days to override the veto by a two-thirds majority vote.

Each of the 18 separate city districts has its own council (*Rada dzielnicy*).[70] Their duties are focused on aiding the President and the City Council, as well as supervising various municipal companies, city-owned property and schools. The head of each of the District Councils is named the Mayor (*Burmistrz*) and is elected by the local council from the candidates proposed by the President of Warsaw.

The mayor of Warsaw is called President. Generally, in Poland, the mayors of bigger cities are called *presidents* – i.e. such cities, which have over 100,000 people or these, where already was president before 1990. The first Warsaw President was Jan Andrzej Menich (1695–1696).[72] Between 1975 and 1990 the Warsaw Presidents was simultaneously the Warsaw Voivode. Since 1990 the President of Warsaw had been elected by the City council.[73] In the years of 1994–1999 the mayor of the district Centrum automatically was designed as the President of Warsaw: the mayor of Centrum was elected by the district council of Centrum and the council was elected only by the Centrum residents. Since 2002 the President of Warsaw is elected by all of the citizens of Warsaw.[73]

The current President of Warsaw is Hanna Gronkiewicz-Waltz (since 2006-12-02) – the former president of the National Bank of Poland.[74] The first president elected according these rules was Lech Kaczyński. When he was elected on the President of Polish Republic (December 2005), there was not an additional election in Warsaw, hence formally he was simultaneously the President of Poland and the President of Warsaw.

Politics

As the capital of Poland, Warsaw is the political centre of the country. All state agencies are located there, including the Polish Parliament, the Presidential Office and the Supreme Court. In the Polish parliament the city and the area are represented by 31 MPs (out of 460). Additionally, Warsaw elects two MEPs.

The Sejm is the lower house of the Polish parliament. The Sejm is made up of 460 deputies, or Poseł in Polish (literally 'Envoy'). It is elected by universal ballot and is presided over by a speaker called the Marshal of the Sejm (Marszałek Sejmu).

The Sejm building.

Transport

Warsaw has seen major infrastructural changes over the past few years amidst increased foreign investment and economic growth. The city has a much improved infrastructure with new roads, flyovers, bridges, etc.[75]

The city centre – view on Dmowski Roundabout

Warsaw lacks a good circular road system and most traffic goes directly through the city centre. Warsaw ring road has been planned consisting of three express roads: S2, S8 and S17. Currently parts of S2 and S8 are under construction and to be completed up to 2012. However thanks to the A2 motorway stretching west from Warsaw which opened in June 2012, the city now has a direct motorway connection with Łódż, Poznań and ultimately with Berlin. The city has two international airports, Warsaw Frédéric Chopin Airport, located just 10 kilometres (**unknown operator: u'strong'** mi) from the city centre, with Warsaw-Modlin Airport located 35 kilometres (**unknown operator: u'strong'** mi) to the north is the city's second international airport in July 2012.[76] With around 100 international and domestic flights a day and with over 9,268,551 passengers served in 2007, Warsaw Frédéric Chopin Airport is by far the biggest airport in Poland.[76]

Public transport in Warsaw includes buses, trams (streetcars), Metro, light rail Warszawska Kolej Dojazdowa line, urban railway Szybka Kolej Miejska, regional rail Koleje Mazowieckie (Mazovian Railways),[77] and bicycle sharing systems (Veturilo and Bemowo Bike). The buses, trams, urban railway and Metro are managed by Zarząd Transportu Miejskiego (*ZTM*, the Warsaw Transport Authority).

The regional rail and light rail is operated by Polish State Railways (PKP). There are also some suburban bus lines run by private operators.[78] Bus service covers the entire city, with approximately 170 routes totalling about 2603 kilometres (**unknown operator: u'strong'** mi) long, and with some 1,600 vehicles.

Plac Wilsona metro station

Currently, the *Tramwaje Warszawskie* (Warsaw Trams) company runs 863 cars on over 240 kilometres (**unknown operator: u'strong'** mi) of tracks. Twenty-odd lines run across the city with additional lines opened on special occasions (such as All Saints' Day).

The first section of the Warsaw Metro was opened in 1995 initially with a total of 11 stations.[79] It now has 21 stations running a distance of approximately 23 kilometres.[80] Initially, all of the trains were Russian built. In 1998, 108 new carriages were ordered from Alstom.[79] The second line running east-west will be about 31 kilometres. The central section is now under construction and will be 6 km. long with seven stations.[79] The main railway station is Warszawa Centralna serving both domestic traffic to almost every major city in Poland, and international connections. There are also five other major railway stations and a number of smaller suburban stations.

Infrastructure

Like many cities in Central and Eastern Europe, infrastructure in Warsaw suffered considerably during communism. However, over the past decade it has seen many improvements due to solid economic growth, an increase in foreign investment as well as funding from the European Union. In particular, the city's underground transit system, roads, sidewalks, health care facilities and sanitation facilities have improved markedly.[75]

Visualization of the National Temple of Divine Providence (under construction).

Today, Warsaw has some of the best medical facilities in Poland and Central Europe. The city is home to the Children's Memorial Health Institute (CMHI), the highest-reference hospital in all of Poland, as well as an active research and education center.[81] While the Maria Skłodowska-Curie Institute of Oncology it is one of the largest and most modern oncological institutions in Europe.[82] The clinical section is located in a 10-floor building with 700 beds, 10 operating theatres, an intensive care unit, several diagnostic departments as well as an outpatient clinic.[82]

Religion

Throughout its existence, Warsaw has been a multi-cultural city.[83] According to a census of 1901, out of 711,988 inhabitants there were 56,2% Catholics, 35,7% Jews, 5% Greek orthodox Christians and 2,8% Protestants.[84] Eight years later, in 1909, there were 281,754 Jews (36,9%), 18,189 Protestants (2,4%) and 2,818 Mariavites (0,4%).[85] This led to construction of hundreds of places of religious worship in all parts of the town. Most of them were destroyed in the aftermath of the Warsaw Uprising of 1944. After the war the new communist authorities of Poland discouraged church construction and only a small part of them were rebuilt.[86]

Leisure activities

Events

Several commemorative events take place every year. Gatherings of thousands of people on the banks of the Vistula on Midsummer's Night for a festival called Wianki (Polish for *Wreaths*) have become a tradition and a yearly event in the programme of cultural events in Warsaw.[87] [88] The festival traces its roots to a peaceful pagan ritual where maidens would float their wreaths of herbs on the water to predict when they would be married, and to whom.[87] By the 19th century this tradition had become a festive event, and it continues today.[87] The city council organize concerts and other events.[88] Each Midsummer's Eve, apart from the official floating of wreaths, jumping over fires, looking for the fern flower, there are musical performances, dignitaries' speeches, fairs and fireworks by the river bank.[88]

Jazz concert in Tygmont Club.

Warsaw Multimedia Fountain Park opened in 2011

The Warsaw Film Festival, an annual festival that takes place every October.[89] Films are usually screened in their original language with Polish subtitles and participating cinemas include Kinoteka (Palace of Science and Culture), Multikino at Golden Terraces and Kultura. Over 100 films are shown throughout the festival, and awards are given to the best and most popular films.[89]

Sports

On 9 April 2008 the President of Warsaw, Hanna Gronkiewicz-Waltz, obtained from the mayor of Stuttgart Wolfgang Schuster a challenge award – a commemorative plaque awarded to Warsaw as the European capital of Sport in 2008.[90]

The National Stadium, capacity of 58,500 seat football (soccer) stadium, Warsaw's recently demolished 10th-Anniversary Stadium.[91] The national stadium hosted the opening match, 2 group matches, a quarterfinal, and a semifinal of the UEFA Euro 2012 hosted jointly by Poland and Ukraine.[92]

There are many sports centres in the city as well. Most of these facilities are swimming pools and sports halls, many of them built by the municipality in the past several years. The main indoor venue is Hala Torwar, used for all kinds of indoor sports (mainly, indoor skating rink). There is also open-air skating rink (Stegny) and the horse racetrack (Służewiec).

The Olympic Center in Warsaw

The best of the city's swimming centres is at Wodny Park Warszawianka, 4 km south of the centre at Merliniego Street, where there's an Olympic-sized pool as well as water slides and children's areas.[93]

From the Warsovian football teams, the most famous is Legia Warszawa – the army club with a nationwide following play at Polish Army Stadium, just southeast of the centre at Łazienkowska Street. Established in 1916, they have won the country's championship 8 times (most recently in 2006) and won the Polish Cup 14 times. They have never been relegated divisions. In the Champions League season 1995/96 they reached the quarter-finals, where they lost to Panathinaikos Athens.

Their local rivals, Polonia Warsaw, have significantly fewer supporters, yet they managed to win Ekstraklasa Championship in 2000. They also won the country's championship in 1946, and won the cup twice as well. Polonia's home venue is located at Konwiktorska Street, a ten-minute walk north from the Old Town.

Culture

Theatre in the past

From 1833 to the outbreak of World War II, Plac Teatralny *(Theatre Square)* was the country's cultural hub and home to the various theatres.[94]

Great Theater, home of Poland's National Theatre and Opera.

The main building housed the Great Theatre from 1833 to 1834, the Rozmaitości Theatre from 1836 to 1924 and then the National Theatre, the Reduta Theatre from 1919 to 1924, and from 1928 to 1939 – the Nowy Theatre, which staged productions of contemporary poetical drama, including those directed by Leon Schiller.[94]

Nearby, in Ogród Saski (the Saxon Garden), the Summer Theatre was in operation from 1870 to 1939,[95] and in the inter-war period, the theatre complex also included Momus, Warsaw's first literary cabaret, and Leon Schiller's musical theatre Melodram. The Wojciech Bogusławski Theatre (1922–26), was the best example of "Polish monumental theatre". From the mid-1930s, the Great Theatre building housed the Upati Institute of Dramatic Arts – the first state-run academy of dramatic art, with an acting department and a stage directing department.[94]

Plac Teatralny and its environs was the venue for numerous parades, celebrations of state holidays, carnival balls and concerts.

Theatre

Warsaw is home to over 30 major theatres spread throughout the city, including the National Theatre (founded in 1765) and the Grand Theatre (established 1778).[96]

Palace of Culture and Science.

Warsaw also attracts many young and off-stream directors and performers who add to the city's theatrical culture. Their productions may be viewed mostly in smaller theatres and *Houses of Culture* (*Domy Kultury*), mostly outside *Śródmieście* (Central Warsaw). Warsaw hosts the International Theatrical Meetings.

Music

Thanks to numerous musical venues, including the Teatr Wielki, the Polish National Opera, the Chamber Opera, the National Philharmonic Hall and the National Theatre, as well as the Roma and Buffo music theatres and the Congress Hall in the Palace of Culture and Science, Warsaw hosts many events and festivals. Among the events worth particular attention are: the International Frederick Chopin Piano Competition, the International Contemporary Music Festival Warsaw Autumn, the Jazz Jamboree, Warsaw Summer Jazz Days, the International Stanisław Moniuszko Vocal Competition, the Mozart Festival, and the Festival of Old Music.[97]

Museums and art galleries

The levelling of Warsaw during the war has left gaping holes in the city's historic collections. [98] Although a considerable number of treasures were spirited away to safety in 1939, a great number of collections from palaces and museums in the countryside were brought to Warsaw at that time as the capital was considered a safer place than some remote castle in the borderlands.[98] Thus losses were heavy.[98]

Warsaw Uprising Museum.

As interesting examples of expositions the most notable are: the world's first Museum of Posters boasting one of the largest collections of art posters in the world,[99] Museum of Hunting and Riding and the Railway Museum. From among Warsaw's 60 museums, the most prestigious ones are National Museum with a collection of works whose origin ranges in time from antiquity till the present epoch as well as one of the best collections of paintings in the country including some paintings from Adolf Hitler's private collection,[100] and Museum of the Polish Army whose set portrays the history of arms.

The collections of Łazienki and Wilanów palaces (both buildings came through the war in good shape) focuses on the paintings of the "old masters", as are those of the Royal Castle which displays the Lanckoroński Collection including two paintings by Rembrandt.[101] The Palace in Natolin, a former rural residence of Duke Czartoryski, is another venue with its interiors and park accessible to tourists.

Holding Poland's largest private collection of art, the Carroll Porczyński Collection Museum[102] displays works from such varied artists as Paris Bordone, Cornelis van Haarlem, José de Ribera, William-Adolphe Bouguereau, Pierre-Auguste Renoir and Vincent van Gogh[103] along with some copies of masterpieces of European painting.

17th century Ostrogski Castle houses the Chopin Museum.

A fine tribute to the fall of Warsaw and history of Poland can be found in the Warsaw Uprising Museum and in the Katyń Museum which preserves the memory of the crime.[104] Museum of Independence host sentimental and patriotic paraphernalia connected with these epochs. Dating back to 1936 Warsaw Historical Museum contains 60 rooms which host a permanent exhibition of the history of Warsaw from its origins until today.

The 17th century Royal Ujazdów Castle currently houses Centre for Contemporary Art, with some permanent and temporary exhibitions, concerts, shows and creative workshops. The Centre currently realizes about 500 projects a year. Zachęta National Gallery of Art, the oldest exhibition site in Warsaw, with a tradition stretching back to the mid-19th century organises exhibitions of modern art by Polish and international artists and promotes art in many other ways.

The city also possesses some oddities such as the Museum of Caricature[105] and a Motorisation Museum in Otrębusy.[106] The newly established in 2011 Erotic Museum in Warsaw is the only such museum in Poland and one of the few of a kind in Europe.[107] Over 2,000 exhibits collected by the Erotic Museum, reveals the erotic fascinations of artists from all continents.[107] Museum collections include the 14th century German chastity belts, erotic Peruvian pottery, Persian miniatures, Tibetan votive paintings, Indian carvings, Chinese porcelain, African folk art, Japanese Shunga paintings, Thai votive penises and many other exhibits.[107]

Media and film

Main TVP headquarters at Woronicza street.

Warsaw is the media centre of Poland, and the location of the main headquarters of TVP and other numerous local and national TV and radio stations, such as TVN, Polsat, TV4, TV Puls, Canal+ Poland, Cyfra+ and MTV Poland.[108]

Since May 1661 the first Polish newspaper, Polish Ordinary Mercury, was printed in Warsaw. The city is also the printing capital of Poland with a wide variety of domestic and foreign periodicals expressing diverse views, and domestic newspapers are extremely competitive. *Rzeczpospolita*, *Gazeta Wyborcza*, *Dziennik Polska-Europa-Świat* Poland's large nationwide daily newspapers[109] have their headquarters in Warsaw.

Warsaw also has a sizable movie and television industry. The city houses several movie companies and studios. Among the movie companies are TOR, Czołówka, Zebra and Kadr who is behind several international movie productions.[110]

Over the next few years the new Film City in Nowe Miasto, located a mere 80 km from Warsaw, will become the centre of Polish film production and international co-production.[110] It is to be the largest high-tech film studio in Europe.[110] The first projects filmed in the new Film City will be two films about the Warsaw Uprising.[110] Two backlots will be constructed for these projects – a lot of pre-WWII Warsaw and city ruins.[110]

Since World War II, Warsaw has been the most important centre of film production in Poland. It has also been featured in numerous movies, both Polish and foreign, for example: *Kanał* and *Korczak* by Andrzej Wajda, *The Decalogue* by Krzysztof Kieślowski, also including Oscar winner *The Pianist* by Roman Polański.[111]

Education

Rococo Czapski Palace houses the Academy of Fine Arts.

Warsaw holds some of the finest institutions of higher education in Poland. It is home to four major universities and over 62 smaller schools of higher education.[112] The overall number of students of all grades of education in Warsaw is almost 500,000 (29.2% of the city population; 2002). The number of university students is over 280,000.[113] Most of the reputable universities are public, but in recent years there has also been an upsurge in the number of private universities.

The University of Warsaw was established in 1816, when the partitions of Poland separated Warsaw from the oldest and most influential Polish academic center, in Kraków.[114] Warsaw University of Technology is the second academic school of technology in the country, and one of the largest in Central Europe, employing 2,000 professors.[115] Other institutions for higher education include the Medical University of Warsaw, the largest medical school in Poland and one of the most prestigious, the National Defence University, highest military academic institution in Poland, the Fryderyk Chopin Music Academy the oldest and largest music school in Poland, and one of the largest in Europe,[116] the Warsaw School of Economics, the oldest and most renowned economic university in the country,[117] and the Warsaw University of Life Sciences the largest agricultural university founded in 1818.[118]

Warsaw University Library

Warsaw has numerous libraries, many of which contain vast collections of historic documents. The most important library in terms of historic document collections include the National Library of Poland. Library holds 8.2 million volumes in its collection.[119] Formed in 1928[120] sees itself as a successor to the Załuski Library, the biggest in Poland and one of the first and biggest libraries in the world.[120] [121]

Another important library – the University Library, founded in 1816,[122] is home to over two million items.[123] The building was designed by architects Marek Budzyński and Zbigniew Badowski and opened on 15 December 1999.[124] It is surrounded by green. The University Library garden, designed by Irena Bajerska, was opened on 12 June 2002. It is one of the largest and most beautiful roof gardens in Europe with an area of more than 10000 m^2 (107639.10 sq ft), and plants covering 5111 m^2 (55014.35 sq ft).[125] As the university garden it is open to the public every day.[125]

Economy

In 2011, Warsaw was ranked the world's 46th most expensive city to live in.[126] It was classified as an Alpha- world city (also known as a "major world city") by the Globalization and World Cities (GaWC) Study Group and Network from Loughborough University, placing it on a par with cities such as Amsterdam or Rome.[127] The city also ranked 8th out of 65 cities on Mastercard's Emerging Markets Index (2008).[128]

Business and commerce

Warsaw city centre as seen from the Vistula

Warsaw, especially its city centre (*Śródmieście*), is home not only to many national institutions and government agencies, but also to many domestic and international companies. In 2006, 304,016 companies were registered in the city.[129] Warsaw's ever-growing business community has been noticed globally, regionally, and nationally. Mastercard Emerging Market Index has noted Warsaw's economic strength and commercial center. Moreover, Warsaw was ranked as the 7th greatest emerging market. Foreign investors' financial participation in the city's development was estimated in 2002 at over 650 million euro. Warsaw produces 12% of Poland's national income,[130] which in 2008 was 305.1% of the Polish average, per capita (or 160% of the European Union average). The GDP per capita in Warsaw amounted to PLN 94 000 in 2008 (ca. EUR 23 800, USD 33 000).[131] Total nominal GDP of the city in 2010 amounted to 191.766 billion PLN, 111696 PLN per capita, which was 301,1 % of Polish average. Warsaw leads the region of Central Europe in foreign investment and in 2006, GDP growth met expectations with a level of 6.1%.[132] It also has one of the fastest growing economies, with GDP growth at 6.5 percent in 2007 and 6.1 percent in the first quarter of 2008.[133]

Golden Terraces shopping centre.

At the same time the unemployment rate is one of the lowest in Poland, not exceeding 3%, according to the official figures. The city itself collects around 8,740,882,000 złotys in taxes and direct government grants.

Warsaw Stock Exchange

Warsaw's first stock exchange was established in 1817 and continued trading until World War II. It was re-established in April 1991, following the end of the post-war communist control of the country and the reintroduction of a free-market economy.[134] Today, the Warsaw Stock Exchange (WSE) is, according to many indicators,[133] the largest market in the region, with 374 companies listed and total capitalization of 162 584 mln EUR as of 31 August 2009.[135] From 1991 until 2000, the stock exchange was, ironically, located in the building previously used as the headquarters of the Polish United Workers' Party (PZPR).[136] The city is considered to be one of the most attractive business locations in Europe.[133]

Exchange Building, home of the exchange from 1876 until World War II.

Industry

During Warsaw's reconstruction after World War II, the communist authorities decided that the city would become a major industrial centre. As a result, numerous large factories were built in and around the city. The largest were the *Huta Warszawa* Steel Works, the car factory FSO and the tractor factory "Ursus".

As the communist economy deteriorated, these factories lost significance and most went bankrupt after 1989.[137][138] Today, the Arcelor Warszawa Steel Mill (formerly *Huta Warszawa*) is the only major factory remaining.

The FSO Car Factory was established in 1951. A number of vehicles have been assembled there over the decades, including the Warszawa, Syrena, Fiat 125p (under license from Fiat, later renamed FSO 125p when the license expired) and the Polonez. The last two models listed were also sent abroad and assembled in a number of other countries, including Egypt and Columbia. In 1995 the factory was purchased by the South Korean car manufacturer Daewoo, which assembled the Tico, Espero, Nubia, Tacuma, Leganza, Lanos and Matiz there for the European market. In 2005 the factory was sold to AvtoZAZ, a Ukrainian car manufacturer which assembled there the Chevrolet Aveo. The license for the production of the Aveo expired in February 2011 and has since not been renewed.

The "Ursus" factory opened in 1893 and is still in operation today. Throughout its history various machinery was assembled there, including motorcycles, military vehicles, trucks and buses. However, since World War II only tractors are still being assembled there.

The number of state-owned enterprises continues to decrease while the number of companies operating with foreign capital is on the rise, reflecting the continued shift towards a modern market-based economy.[137] The largest foreign investors are Daewoo, Coca-Cola Amatil and Metro AG.[137] Warsaw has the biggest concentration of electronics and high-tech industry in Poland, while the growing consumer market perfectly fosters the development of the food-processing industry.[137]

Tourist attractions

Sights

Historic Centre of Warsaw	
Name as inscribed on the World Heritage List [139]	

Country	Poland
Type	Cultural
Criteria	ii, vi
Reference	30 [140]
UNESCO region [141]	Europe
Inscription history	
Inscription	1980 (4th Session)

Although today's Warsaw is a fairly young city, it has many tourist attractions. Apart from the Warsaw Old Town quarter, reconstructed after World War II, each borough has something to offer. Among the most notable landmarks of the Old Town are the Royal Castle, King Zygmunt's Column, Market Square, and the Barbican.

Further south is the so-called Royal Route, with many classicist palaces, the Presidential Palace and the University of Warsaw campus. Wilanów Palace, the former royal residence of King John III Sobieski, is notable for its baroque architecture and parks.[142]

Warsaw's oldest public park, the Saxon Garden, is located within 10 minutes' walk from the old town.[143] Warsaw's biggest public park is the Royal Baths Park, established in the 17th century and given its current classical shape in late 18th century.[144] It is located further south, on the Royal Route, about 3 km (**unknown operator: u'strong'** mi) from the Warsaw Old Town.

The Powązki Cemetery is one of the oldest cemeteries in Europe,[145] full of sculptures, some of them by the most renowned Polish artists of the 19th and 20th centuries. Since it serves the religious communities of Warsaw, be it Catholics, Orthodox, Jews, Muslims or Protestants, it is often called a necropolis. Nearby is the Okopowa Street Jewish Cemetery, one of the largest Jewish cemeteries in Europe.

In many places in the city the Jewish culture and history resonates down through time.[146] Among them the most notable are the Jewish theater, the Nożyk Synagogue, Janusz Korczak's Orphanage and the picturesque Próżna Street.[146] The tragic pages of Warsaw's history are commemorated in places such as the Monument to the Ghetto Heroes, the Umschlagplatz, fragments of the Ghetto wall on Sienna Street and a mound in memory of the Jewish Combat Organization.[146]

There are also many places commemorating the heroic history of Warsaw.[147] Pawiak, an infamous German Gestapo prison now occupied by a Mausoleum of Memory of Martyrdom and the museum, is only the beginning of a walk in the traces of Heroic City.[147] The Warsaw Citadel, an impressive 19th century fortification built after the defeat of the November Uprising, was a place of martyr for the Poles.[147] Another important monument, the statue of Little Insurgent located at the ramparts of the Old Town, commemorates the children who served as messengers and frontline troops in the Warsaw Uprising, while the impressive Warsaw Uprising Monument by Wincenty Kućma was erected in memory of the largest insurrection of World War II.[147] [148]

In Warsaw there are many places connected with the life and work of Frédéric Chopin. The heart of Polish-born composer is sealed inside Warsaw's Holy Cross Church.[149] During the summer time the Chopin Statue in the Royal Baths Park is a place where pianists give concerts to the park audience.[150]

Also many references to Marie Curie, her work and her family can be found in Warsaw: Marie's birthplace at the Warsaw New Town, the working places where she did her first scientific works[151] and the Radium Institute at Wawelska Street for the research and the treatment of cancer which she founded in 1925.[152]

Warsaw Mermaid

The mermaid (*syrenka*) is Warsaw's symbol[153] and can be found on statues throughout the city and on the city's coat of arms. This imagery has been in use since at least the mid-14th century.[154] The oldest existing armed seal of Warsaw is from the year 1390, consisting of a round seal bordered with the Latin inscription *Sigilium Civitatis Varsoviensis* (Seal of the city of Warsaw).[155] City records as far back as 1609 document the use of a crude form of a sea monster with a female upper body and holding a sword in its claws.[156] In 1653 the poet Zygmunt Laukowski asks the question:

The 1659 coat of arms of Old Warsaw on the cover of one of Warsaw's accounting books.

> *Warsaw of strong walls; why was the emblem Mermaid with sharp sword, given you by the kings?*
>
> —Zygmunt Laukowski[157]

1855 bronze sculpture of *The Warsaw Mermaid* in the Old Town Market Place

The origin of the legendary figure is not fully known. The best-known legend, by Artur Oppman, is that long ago two of Triton's daughters set out on a journey through the depths of the oceans and seas. One of them decided to stay on the coast of Denmark and can be seen sitting at the entrance to the port of Copenhagen. The second mermaid reached the mouth of the Vistula River and plunged into its waters. She stopped to rest on a sandy beach by the village of Warszowa, where fishermen came to admire her beauty and listen to her beautiful voice. A greedy merchant also heard her songs; he followed the fishermen and captured the mermaid.[158]

Another legend says that a mermaid once swam to Warsaw from the Baltic Sea for the love of the Griffin, the ancient defender of the city, who was killed in a struggle against the Swedish invasions of the 17th century. The mermaid, wishing to avenge his death, took the position of defender of Warsaw, becoming the symbol of the city.[158]

Every member of the Queen's Royal Hussars of the United Kingdom light cavalry wears the *Maid of Warsaw*, the crest of the City of Warsaw, on the left sleeve of his No. 2 (Service) Dress.[159] Members of 651 Squadron Army Air Corps of the United Kingdom also wear the *Maid of Warsaw* on the left sleeve of their No. 2 (Service) Dress.[160]

Famous people

One of the most famous people born in Warsaw was Maria Skłodowska-Curie, who achieved international recognition for her research on radioactivity.[161] Famous musicians include Władysław Szpilman and Frédéric Chopin. Chopin was born in the village of Żelazowa Wola, about 60 kilometers from Warsaw, but moved to the city with his family when he was seven months old.[162] Kazimierz Pułaski, a hero of the American Revolutionary War, was born here in 1745.

Maria Skłodowska-Curie was born in Warsaw.

Tamara de Lempicka was a famous artist born in Warsaw.[163] She was born Maria Górska in Warsaw to wealthy parents and in 1916 married a Polish lawyer Tadeusz Łempicki.[164] Better than anyone else she represents the Art Deco style in painting.[163] Nathan Alterman, the Israeli poet, was born in Warsaw, as was Moshe Vilenski, the Israeli composer, lyricist, and pianist, who studied music at the Warsaw Conservatory.[165] Warsaw was the beloved city of Isaac Bashevis Singer, which he described in many of his novels:[166] *Warsaw has just now been destroyed. No one will ever see the Warsaw I knew. Let me just write about it. Let this Warsaw not not disappear forever*, he commented.[167]

Rankings

- Largest cities of the European Union: ranked 6th (2012).
- Most expensive cities: ranked 113th of 144.
- Best cities in terms of quality of living: ranked 87th of 100.

International relations

Twin towns and sister cities

Warsaw is twinned with:[168]

- Astana in Kazakhstan *(since 2002)*
- Bangkok in Thailand *(since 2010)*
- Berlin in Germany *(since 1991)*[169]
- Budapest in Hungary *(since 2005)*
- Buenos Aires, Argentina *(since 1992)*[170]
- Coventry in the United Kingdom *(since 1957)*
- Chicago in the United States *(since 1960)*
- Düsseldorf in Germany *(since 1989)*[171]
- Grozny in Russia *(since 1997)*
- The Hague in Netherlands *(since 1991)*
- Hamamatsu in Japan *(since 1990)*
- Hanoi in Vietnam *(since 2000)*
- Harbin in China *(since 1993)*
- Île-de-France in France *(since 1990)*
- Istanbul in Turkey *(since 1991)*[172] [173]
- Kiev in Ukraine *(since 1994)*
- Madrid in Spain *(since 1981)*[174]
- Moscow in Russia *(since 1993)*
- Oslo in Norway *(since 2005)*[175]
- Paris in France *(since 1999)*
- Rio de Janeiro in Brazil *(since 1997)*
- Riga in Latvia *(since 2002)*[176]
- Saint-Étienne in France *(since 1995)*
- St. Petersburg in Russia *(since 1997)*[177]
- San Diego in the United States *(since 1960)*[178]
- Seoul in South Korea *(since 1996)*[179]
- Taipei in Taiwan *(since 1995)*[180]
- Tel Aviv in Israel *(since 1992)*[181]
- Toronto in Canada *(since 1990)*
- Vienna in Austria *(since 2001)*
- Vilnius in Lithuania *(since 1998)*

References – city's official site.[182]

See also

- Tourism in Poland
- Battle of Warsaw
- Street names of Warsaw
- Warsaw dialect
- Warsaw Fire Brigade
- Warsaw concentration camp

References

Bibliography

- Crowley, David (2003). *Warsaw* [183]. Reaktion Books. ISBN 1-86189-179-2. Retrieved 28 August 2011.
- Olchowik-Adamowska, Liliana; Ławecki, Tomasz (1 April 2006). *Travellers Warsaw* [184] (First ed.). Peterborough, United Kingdom: Thomas Cook Publishing. ISBN 978-1-84157-492-9. Retrieved 11 March 2010.

Notes

[1] (Polish) "Ludność w gminach według stanu w dniu 31.12.2011 r. bilans opracowany w oparciu o wyniki NSP'2011" (http://www.stat.gov.pl/gus/5840_13211_PLK_HTML.htm). Glówny Urząd Statystyczny. . Retrieved 21 August 2012.

[2] http://www.um.warszawa.pl

[3] "European Metropolitan Transport Authorities" (http://www.emta.com/article.php3?id_article=43). EMTA. . Retrieved 3 June 2011.

[4] "CityProfiles" (http://www.urbanaudit.org/CityProfiles.aspx?CityCode=PL001C&CountryCode=PL). Urban Audit. . Retrieved 3 June 2011.

[5] (English) "Warsaw" (http://web.archive.org/web/20080603015458/http://www.goeuro2012.com/html/warsaw.html). *www.goeuro2012.com*. Archived from the original (http://www.goeuro2012.com/html/warsaw.html) on 3 June 2008. . Retrieved 15 July 2008.

[6] "Top 150 City Destinations: London Leads the Way" (http://blog.euromonitor.com/2007/10/top-150-city-destinations-london-leads-the-way.html). blog.euromonitor.com. 11 October 2007. . Retrieved 18 November 2011. "City Ranking '000 tourist arrivals / Warsaw 29 2,925"

[7] George Friedman (30 November 2010). "Geopolitical Journey, Part 6: Ukraine" (http://www.stratfor.com/weekly/20101129_geopolitical_journey_part_6_ukraine). www.stratfor.com. . Retrieved 18 November 2011. "Far more interesting than their

geopolitical speculation was their fixation on Warsaw. (...) But what was most interesting was how little talk there was of Ukrainian oligarchs compared to Warsaw markets. The oligarchs might have been way beyond them and therefore irrelevant, but it was Warsaw, not the European Union or the power structure, that got their juices flowing."

[8] "Warsaw Stock Exchange to List Bulgarian Companies" (http://www.novinite.com/view_news.php?id=113177). www.novinite.com. 15 February 2010. . Retrieved 18 November 2011. "Bulgarian public companies will now be able to secure the financing of their projects through investors on the Warsaw Stock Exchange, according to the sitẻ DarikFinance."

[9] (English) "The SETAC Europe 18th Annual Meeting" (http://web.archive.org/web/20080528233643/http://www.setac.eu/warsaw/city.html). *www.setac.eu*. Archived from the original (http://www.setac.eu/warsaw/city.html) on 28 May 2008. . Retrieved 22 January 2009.

[10] (Polish) "The city of phoenix – War*saw everything" (http://miastofeniksa.pl/). . Retrieved 22 January 2009.

[11] (English) "Coat of Arms and Colours of the Capital City of Warsaw" (http://bip.warszawa.pl/English/Main_Menu/capital_city_of_warsaw/coat_of_arms_colours.htm). *bip.warszawa.pl*. . Retrieved 14 January 2009.

[12] Czerkawski, Andrzej; Jurga, Tadeusz (1969). *Dla ciebie ojczyzno* (http://books.google.pl/books?ei=O7XkUM67IZDR4QTbyYHICQ&hl=pl&id=dM8qAQAAIAAJ&dq=Capital+City+of+Warsaw+1940+To+the+inhabitants+of+the+Capital+City+of+Warsaw+â + in+recognition+of+their+heroism+and&q=1940). Sport i Turystyka. p. 435. . "ORDER OF VALOUR "VIRTUTI MILITARI", FIFTH CLASS Capital City of Warsaw 1940 To the inhabitants of the Capital City of Warsaw — in recognition of their heroism and unshakable bravery in the struggle with the Nazi aggressor."

[13] (Polish) Piotr Olechno (28 August 2008). "Sen zamiast Warszawianki" (http://www.polskatimes.pl/warszawa/40061,sen-zamiast-warszawianki,id,t.html). *www.polskatimes.pl*. . Retrieved 3 March 2009.

[14] (Polish) Kazimierz Rymut (1987). *Nazwy miast Polski*. Zakład Narodowy im. Ossolińskich. ISBN 83-04-02436-5.

[15] "The Warsaw Mermaid" (http://www.warsaw-life.com/poland/warsaw-legend). . Retrieved 11 February 2008.

[16] "Historia Warszawy" (http://web.archive.org/web/20100513082637/http://www.um.warszawa.pl/v_syrenka/miasto/historia.htm) (in Polish). Archived from the original (http://www.um.warszawa.pl/v_syrenka/miasto/historia.htm) on 13 May 2010. . Retrieved 11 February 2008.

[17] (Polish) "Ustawa o ustroju miasta stołecznego Warszawy" (http://web.archive.org/web/20070101150852/http://www.prawo.lex.pl/bap/samorzad/Dz.U.2002.41.361.html). *www.prawo.lex.pl*. Archived from the original (http://www.prawo.lex.pl/bap/samorzad/Dz.U.2002.41.361.html) on 1 January 2007. . Retrieved 15 July 2008.

[18] (Polish) *Historia Warszawy (History of Warsaw)*. Warsaw. 2004. ISBN 83-89632-04-7.

[19] (English) "Warsaw's history" (http://www.e-warsaw.pl/miasto/historia.htm). *www.e-warsaw.pl*. . Retrieved 24 July 2008.

[20] (English) Neal Ascheron. "The Struggles for Poland" (http://www.halat.pl/poland2.html). *www.halat.pl*. . Retrieved 24 July 2008.

[21] (English) Michał Rożek, Doris Ronowicz (1988). *Cracow: a treasury of Polish culture and art*. Interpress Publishers. p. 74. ISBN 83-223-2245-3.

[22] (English) John Stanley (March–June 2004). "Literary Activities and Attitudes in the Stanislavian Age in Poland (1764–1795): A Social System?" (http://web.archive.org/web/20110514133024/http://findarticles.com/p/articles/mi_qa3763/is_200403/ai_n9363971/?tag=content;col1). *findarticles.com*. Archived from the original (http://findarticles.com/p/articles/mi_qa3763/is_200403/ai_n9363971/?tag=content;col1) on 14 May 2011. . Retrieved 23 April 2009.

[23] (English) Cornelia Golna (2004). *City of man's desire: a novel of Constantinople* (http://books.google.pl/books?id=xHXGa8HSQIQC&printsec=frontcover#v=onepage&q&f=false). Go-Bos Press. p. 318. ISBN 90-804114-4-2. .

[24] (French) Zbigniew Naliwajek. Romain Rolland et la littérature polonaise (http://www.cairn.info/article.php?ID_REVUE=RLC&ID_NUMPUBLIE=RLC_307&ID_ARTICLE=RLC_307_0325). Revue de littérature comparée 3/2003 (n°307), p. 325-338.

[25] (English) Augustin P. O'Brien (1864). *Petersburg and Warsaw: Scenes Witnessed During a Residence in Poland and Russia in 1863–64* (http://books.google.com/?id=WlYBAAAAQAAJ&printsec=frontcover&dq=Petersburg+and+Warsaw:+scenes+witnessed#PPR1,M1). R. Bentley. . Retrieved 28 January 2009.

[26] (English) "Vistula River Victory" (http://web.archive.org/web/20070509034826/http://www.electronicmuseum.ca/Soviet-Polish-War/spw_3.html). *www.electronicmuseum.ca*. Archived from the original (http://www.electronicmuseum.ca/Soviet-Polish-War/spw_3.html) on 9 May 2007. . Retrieved 14 July 2008.

[27] (English) "Poland, History " Poland in the 20th century " From the Treaty of Versailles to the Treaty of Riga" (http://www.britannica.com/EBchecked/topic/466681/Poland/28213/From-the-Treaty-of-Versailles-to-the-Treaty-of-Riga#ref=ref397007&tab=active~checked,items~checked&title=Poland :: From the Treaty of Versailles to the Treaty of Riga -- Britannica Online Encyclopedia). *www.britannica.com*. . Retrieved 14 July 2008.

[28] (English) Zdzisław G. Kowalski. "Documents of the Battle of Warsaw 1920" (http://www.archiwa.gov.pl/memory/sub_listakrajowa/index.php?va_lang=en&fileid=019). *Memory of the World*. . Retrieved 14 July 2008.

[29] (English) M.M. (2 August 2006). "Warsaw: A Last Glimpse" (http://www.warsawvoice.pl/view/11932/). *www.warsawvoice.pl*. . Retrieved 29 July 2008.

[30] (English) "Warsaw" (http://www.ushmm.org/wlc/article.php?lang=en&ModuleId=10005069). *United States Holocaust Memorial Museum*. . Retrieved 29 July 2008.

[31] Snyder (2010). *Bloodlands*. London: The Bodley Head. p. 280 (http://books.google.co.uk/books?id=BQ1HKmG9xZ8C&pg=PA280).

[32] (English) "The Warsaw Ghetto Uprising" (http://web.archive.org/web/20080517043736/http://www.ushmm.org/outreach/wgupris.htm). *United States Holocaust Memorial Museum*. Archived from the original (http://www.ushmm.org/outreach/wgupris.htm) on 17 May

2008. . Retrieved 29 July 2008.
[33] (English) "The Warsaw Ghetto Uprising" (http://web.archive.org/web/20080623220055/http://www.aish.com/holocaust/overview/he05n27.htm). *www.aish.com*. Archived from the original (http://www.aish.com/holocaust/overview/he05n27.htm) on 23 June 2008. . Retrieved 29 July 2008.
[34] (English) "Warsaw Uprising of 1944" (http://www.warsawuprising.com/timeline.htm). *www.warsawuprising.com*. . Retrieved 14 July 2008.
[35] *The Soviet troops, ordered by Stalin to wait until the Germans had destroyed the remnants of Polish resistance, then moved into what was left of Warsaw, flushed out the remaining Germans, and proclaimed themselves liberators of the city.*
(English) Wesley Adamczyk (2004). *When God looked the other way: an odyssey of war, exile, and redemption* (http://books.google.com/?id=77sneNTXojQC&pg=PA170&dq=The+Soviet+troops,+ordered+by+Stalin+to+wait+until#v=onepage&q=The Soviet troops, ordered by Stalin to wait until&f=false). University of Chicago Press. p. 170. ISBN 0-226-00443-0. .
[36] Borkiewicz, Adam (1957). *Powstanie warszawskie 1944: zarys działań natury wojskowej*. Warsaw: PAX.
[37] (English) "Warsaw Uprising of 1944" (http://www.warsawuprising.com/faq.htm#Warsaw Ghetto Uprising). *www.warsawuprising.com*. . Retrieved 14 July 2008.
[38] (English) "Historic Centre of Warsaw" (http://whc.unesco.org/en/list/30). *whc.unesco.org*. . Retrieved 24 July 2008.
[39] (English) "Pope in Warsaw" (http://www.destinationwarsaw.com/site.php5/Show/135.html). *www.destinationwarsaw.com*. . Retrieved 5 February 2009.
[40] (English) "Attracting foreign investments" (http://www.polandtrade.com.hk/new/eng/news_september2004.htm). *www.polandtrade.com.hk*. The Warsaw Voice. . Retrieved 24 July 2008.
[41] (English) "The National Stadium in Warsaw" (http://www.poland2012.net/stadiums-in-poland/). *www.poland2012.net*. . Retrieved 24 July 2008.
[42] (English) "Geography of Warsaw" (http://geography.howstuffworks.com/europe/geography-of-warsaw.htm). *geography.howstuffworks.com*. . Retrieved 27 February 2009.
[43] http://www.e-warsaw.pl/images/mapa_dzielnice.gif
[44] (Polish) "Dzielnice" (http://web.archive.org/web/20080701114815/http://www.um.warszawa.pl/v_syrenka/miasto/dzielnice.htm). *www.um.warszawa.pl*. Archived from the original (http://www.um.warszawa.pl/v_syrenka/miasto/dzielnice.htm) on 1 July 2008. . Retrieved 11 July 2008.
[45] (Polish) "Pałac Leopolda Kronenberga" (http://www.warszawa1939.pl/index.php?r1=malachowskiego_4&r3=0). *www.warszawa1939.pl*. . Retrieved 29 July 2008.
[46] (English) "A town house of Burbach family" (http://um.warszawa.pl/v_syrenka/perelki/index_en.php?mi_id=43&dz_id=2). *eGuide / Treasures of Warsaw on-line*. . Retrieved 23 February 2009.
[47] (Polish) "Politechnika Warszawska" (http://www.warszawa1939.pl/index.php?r1=politechnika&r3=0). *www.warszawa1939.pl*. . Retrieved 27 February 2009.
[48] (English) "As good as new" (http://www.e-warsaw.pl/new/index.php?dzial=aktualnosci&ak_id=551&kat=3). *www.e-warsaw.pl*. . Retrieved 30 January 2009.
[49] (English) Sampo Ruoppila (2004). *Processes of Residential Differentiation in Socialist Cities* (http://www.nordregio.se/ejsd/refereed9.pdf). European Journal of Spatial Development. pp. 9–10. . Retrieved 10 October 2010.
[50] (English) "Metropolitan Life" (http://www.warsawvoice.pl/view/4682/). *www.warsawvoice.pl*. 4 February 2004. . Retrieved 23 February 2009.
[51] (English) "Europes Top Skyscraper Cities" (http://www.skyscrapernews.com/britains.htm). *www.skyscrapernews.com*. . Retrieved 29 April 2009.
[52] (English) Warsaw Tourist Office. "Parks & Gardens" (http://www.warsawtour.pl/en/warsaw-for-everyone/parks-gardens-2075.html). *www.warsawtour.pl*. . Retrieved 23 February 2009. "Warsaw is a green city. Almost 1/4 of its area is comprisedof fields, parks, green squares and lush gardens, making Warsaw a European metropolis that truly offers its visitors a breath of fresh air."
[53] (Polish) "Parki i lasy Warszawy" (http://www.um.warszawa.pl/v_syrenka/miasto/parki-5.php). *www.um.warszawa.pl*. . Retrieved 25 February 2009.
[54] (Polish) "Nowa Pomarańczarnia" (http://web.archive.org/web/20060208182537/http://um.warszawa.pl/v_syrenka/perelki/?mi_id=148&dz_id=14). *ePrzewodnik / Perełki Warszawy on-line*. Archived from the original (http://um.warszawa.pl/v_syrenka/perelki/?mi_id=148&dz_id=14) on 8 February 2006. . Retrieved 24 February 2009.
[55] (Polish) "Park Praski" (http://web.archive.org/web/20100313065613/http://zielona.um.warszawa.pl/tereny-zielone/park-praski). *zielona.um.warszawa.pl*. Archived from the original (http://zielona.um.warszawa.pl/tereny-zielone/park-praski) on 13 March 2010. . Retrieved 19 April 2011. *Powstał w latach 1865–1871, według projektu Jana Dobrowolskiego, na prawym brzegu Wisły.*
[56] (Polish) "Park Praski" (http://web.archive.org/web/20080314235541/http://warszawa.naszemiasto.pl/rekreacja_i_wypoczynek/58717_63.html). *naszemiasto.pl*. Archived from the original (http://warszawa.naszemiasto.pl/rekreacja_i_wypoczynek/58717_63.html) on 14 March 2008. . Retrieved 18 February 2008.
[57] (English) "Nature reserves as a refuge of Grifola frondosa (DICKS.: FR.) GRAY in central Poland" (http://bpn.com.pl/index.php?option=com_content&task=view&id=230&Itemid=170). *bpn.com.pl*. . Retrieved 24 February 2009.
[58] (English) "Kayaking on the Vistula" (http://www.warsawvoice.pl/view/12179/). *www.warsawvoice.pl*. 30 August 2006. . Retrieved 24 February 2009.

[59] (English) "Warsaw Zoo" (http://www.zoo.waw.pl/). *www.zoo.waw.pl*. . Retrieved 24 February 2009.

[60] Warsaw Zoo opened 11 March 1928, on Ratuszowa Street. It was not the first zoological garden in Warsaw; King Jan Sobieski III kept a court menagerie in Wilanów. Several private zoos were also established in Warsaw in the 19th century. (English) "New Zoo Revue" (http://www.warsawvoice.pl/view/2044/). *www.warsawvoice.pl*. 24 April 2003. . Retrieved 9 May 2009.

[61] (English) Vernon N. Kisling, ed. (2000). *Zoo and aquarium history: ancient animal collections to zoological gardens* (http://books.google.pl/books?id=dxTrR5nOE0UC&printsec=frontcover). CRC Press. pp. 118–119. ISBN 0-8493-2100-X. .

[62] (English) "Demographic Yearbooks of Poland 1939–1979, 1980–1994" (http://www.stat.gov.pl/gus/index_ENG_HTML.htm). *www.stat.gov.pl*. Central Statistical Office of Poland. . Retrieved 29 August 2008.

[63] "Ludność w gminach. Stan w dniu 31 marca 2011 r. - wyniki spisu ludności i mieszkań 2011 r." (http://www.stat.gov.pl/gus/5840_655_PLK_HTML.htm). Główny Urząd Statystyczny. . Retrieved 20 August 2012.

[64] (English) Joshua D. Zimmerman (2004). *Poles, Jews and the politics of nationality* (http://books.google.com/?id=6sbr9cZyw_4C&pg=PA16&dq=population+Brest+Poles+Jews). University of Wisconsin Press. p. 16. ISBN 0-299-19464-7. .

[65] (English) "Warsaw" (http://www.ushmm.org/wlc/article.php?lang=en&ModuleId=10005069). *www.ushmm.org*. . Retrieved 29 August 2008.

[66] (German) F.A. Brockhaus Verlag Leipzig (1935). *Der Grosse Brockhaus: Handbuch des Wissens*. **20** (15 ed.). Brockhaus. p. 25.

[67] (English) Dānishgāh-i Tihrān. Faculty of Fine Arts (1990). *International Conference on Reconstruction of War-Damaged Areas: 6–16 March 1986 : Faculty of Fine Arts, University of Tehran, Iran*. University of Tehran Press. p. 148.

[68] (Polish) Michał Kopiński (28 November 2008). "Warszawa da się lubić? Nie w Poznaniu" (http://poznan.gazeta.pl/poznan/1,36001,6000225.html). *poznan.gazeta.pl/poznan*. . Retrieved 29 September 2010.

[69] (Polish) Joanna Blewąska (28 November 2008). "Warszawa da się lubić?" (http://wyborcza.pl/1,76842,5999966,Warszawa_da_sie_lubic_.html). *wyborcza.pl*. . Retrieved 29 September 2010.

[70] (English) "Administration" (http://e-warsaw.pl/2/index.php?id=568). *e-warsaw.pl*. . Retrieved 31 January 2009.

[71] (English) Uwe Altrock (2006). *Spatial planning and urban development in the new EU member states: from adjustment to reinvention*. Ashgate Publishing, Ltd. p. 262. ISBN 0-7546-4684-X.

[72] (Polish) Barbara Petrozolin-Skowrońska (1994). "Encyklopedia Warszawy" (http://books.google.com/?id=BjjjAAAAMAAJ). *Warsaw Encyclopedia*. Polish Scientific Publishers PWN. p. 94. ISBN 83-01-08836-2. .

[73] (English) Masa Djordjevic (2006). [web.ceu.hu/polsci/ADC/2006/papers/Djordjevic%20Masa%20paper.doc *Politics of Urban Development Planning: Building Urban Governance in Post-Socialist Warsaw?*]. Ashgate Publishing, Ltd. p. 8. web.ceu.hu/polsci/ADC/2006/papers/Djordjevic%20Masa%20paper.doc. Retrieved 10 October 2010.

[74] (English) "Mayor of Warsaw" (http://www.e-warsaw.pl/2/index.php?id=566). *e-warsaw.pl*. . Retrieved 10 October 2010.

[75] (English) Michal Jeziorski (7 March 2007). "Improving Infrastructure" (http://www.warsawvoice.pl/view/14144). *www.warsawvoice.pl*. . Retrieved 28 May 2009.

[76] (English) "Frederick Chopin International Airport" (http://www.airport-technology.com/projects/fredericchipin/). *www.airport-technology.com*. . Retrieved 24 July 2008.

[77] (English) "Public transport" (http://www.e-warsaw.pl/miasto/transport.htm). *www.e-warsaw.pl*. . Retrieved 22 August 2008.

[78] (English) "From monopoly towards market" (http://siteresources.worldbank.org/INTTRANSPORT/Resources/336291-1119275973157/td-ut5.pdf). *siteresources.worldbank.org*. . Retrieved 22 August 2008.

[79] (English) "A History of Subway Construction" (http://web.archive.org/web/20061210054546/http://www.metro.waw.pl/page.php?id=111&PHPSESSID=9eb61f8f4408e4e6260cabb63fc4daa3). *www.metro.waw.pl*. Archived from the original (http://www.metro.waw.pl/page.php?id=111) on 10 December 2006. . Retrieved 30 January 2009.

[80] (English) "Technical and Operating Data of the Existing Subway Section" (http://web.archive.org/web/20070117051410/http://www.metro.waw.pl/page.php?id=56&PHPSESSID=12053adabbe490da1246a5eaaacb78d8). *www.metro.waw.pl*. Archived from the original (http://www.metro.waw.pl/page.php?id=56) on 17 January 2007. . Retrieved 30 January 2009.

[81] (English) Ewa Pronicka and coordinators (27 April 2004). "Perfect for Children" (http://www.warsawvoice.pl/view/5508/). *www.warsawvoice.pl*. . Retrieved 2 March 2009.

[82] (English) Denise Wise, PT, PhD, with Kristin Wodzinski, PT. "People to People: Russia and Poland" (http://web.archive.org/web/20081201112737/http://www.apta.org/AM/Template.cfm?Section=Home&TEMPLATE=/CM/HTMLDisplay.cfm&CONTENTID=28705). *www.apta.org*. Archived from the original (http://www.apta.org/AM/Template.cfm?Section=Home&TEMPLATE=/CM/HTMLDisplay.cfm&CONTENTID=28705) on 1 December 2008. . Retrieved 2 March 2009.

[83] (English) Geert Mak (2008). *In Europe: travels through the twentieth century*. Pantheon Books. p. 427. ISBN 0-307-28057-8. *Today Warsaw is a monocultural city, which is some people's ideal. But before 1939 it was a typically multicultural society. Those were the city's most productive years. We lost that multicultural character during the war.*

[84] (German) Hermann Julius Meyer (1909). *Meyers Konversations-Lexikon*. **20** (6 ed.). Leipzig and Vienna. p. 388.

[85] (German) Erich Zechlin (1916). *Die Bevölkerungs- und Grundbesitzverteilung im Zartum Polen (The distribution of population and property in tsaristic Poland)*. Reimer, Berlin. pp. 82–83.

[86] (English) Marian S. Mazgaj (2010). *Church and State in Communist Poland: A History, 1944–1989*. McFarland. p. 67. ISBN 0-7864-5904-2.

[87] (English) Staś Kmieć. "Midsummer's Eve" (http://www.polamjournal.com/Library/Holidays/Sobotka/sobotka.html). *www.polamjournal.com*. . Retrieved 2 February 2009.

[88] (Polish) Staś Kmieć. "Wianki 2008" (http://web.archive.org/web/20090417070237/http://www.aktivist.pl/wydarzenie/eventId,393787,wianki-2008-wydarzenie.html). *www.aktivist.pl*. Archived from the original (http://www.aktivist.pl/wydarzenie/eventId,393787,wianki-2008-wydarzenie.html) on 17 April 2009. . Retrieved 2 February 2009.
[89] (English) "Warsaw Film Festival" (http://www.wff.pl/en/o-festiwalu/). *www.wff.pl*. . Retrieved 16 February 2009.
[90] (English) "European Capitals of Sport" (http://www.aces-europa.eu/LinkClick.aspx?fileticket=i8gr5Zb1M8I=&tabid=55&mid=379). *www.aces-europa.eu*. . Retrieved 30 January 2009.
[91] (English) Ryan Lucas. "UEFA turns attention to Euro 2012" (http://sports.sportsillustrated.cnn.com/euro/story.asp?i=20080630175055520000101&ref=hea&tm=). *sports.sportsillustrated.cnn.com*. . Retrieved 31 January 2009.
[92] (English) "Warsaw" (http://web.archive.org/web/20080803222406/http://www.e2012.org/en/4_51.html). *www1.e2012.org*. Archived from the original (http://www1.e2012.org/en/4_51.html) on 3 August 2008. . Retrieved 31 January 2009.
[93] (English) "Wodny Park" (http://www.wodnypark.com.pl/index.php?lang=en). *www.wodnypark.com.pl*. . Retrieved 31 January 2009.
[94] (English) "The Theatre's history" (http://web.archive.org/web/20080418103152/http://www.teatrwielki.pl/show_book.php?book=historia&nlang=en). *www.teatrwielki.pl*. 1998. Archived from the original (http://www.teatrwielki.pl/show_book.php?book=historia&nlang=en) on 18 April 2008. . Retrieved 21 February 2008.
[95] (Polish) "Teatr Letni" (http://www.warszawa1939.pl/index.php?r1=letni&r3=0). *warszawa1939.pl*. . Retrieved 14 February 2008.
[96] "Teatr Wielki-Polish National Opera" (http://web.archive.org/web/20080208223208/http://www.teatrwielki.pl/show_book.php?book=historia). Archived from the original (http://www.teatrwielki.pl/show_book.php?book=historia) on 8 February 2008. . Retrieved 11 February 2008.
[97] (English) *Poland* (http://books.google.com/?id=YgQ0B1CNYfQC&pg=PA70&dq=guide+warsaw#PPA128,M1). Rough Guides. 2002. ISBN 1-85828-849-5. .
[98] (Polish) Włodzimierz Kalicki. "Sztuka zagrabiona" (http://www.poloniacal.org/sztuka/sztuka1.htm). Polish American Congress of Southern California. . Retrieved 30 January 2009.
[99] (English) "The Poster Museum at Wilanów" (http://www.postermuseum.pl/en/page/show/history). *www.postermuseum.pl*. . Retrieved 10 February 2009.
[100] Schwarz, Birgit (2009). *Geniewahn: Hitler und die Kunst*. Böhlau Verlag Wien. p. 312. ISBN 32-05783-07-7. "Mehrere Gemälde aus dem Berghof befinden sich heute im Nationalmuseum in Warschau. Bordones Venus und Amor etwa (Abb. 100) ebenso wie der Madonnen-Tondo Bugiardinis (Abb. 62) oder ein großes Ruinenbild von Pannini, das in der verglasten Veranda gehangen hatte (Abb. 113)."
[101] Wetering, van de, Ernst (2005). *A Corpus of Rembrandt Paintings IV: Self-Portraits*. Springer. p. 245. ISBN 14-02032-80-3.
[102] Official name: *Museum of John Paul II Collection*
[103] (English) "Museum of John Paul II Collection" (http://www.muzeummalarstwa.pl/collection.htm). *www.muzeummalarstwa.pl*. . Retrieved 24 February 2009.
[104] (English) Mark Baker, Kit F. Chung (2009). *Frommer's Poland* (http://books.google.pl/books?id=ypJ5fzK_uSkC&printsec=frontcover#v=onepage&q&f=false). Frommer's. p. 79. ISBN 0-470-15819-0. .
[105] (English) "Exhibitions" (http://www.warsaw.com/v/exhibitions/). *www.warsaw.com*. . Retrieved 30 January 2009.
[106] (English) "Museum history" (http://web.archive.org/web/20090129094040/http://muzeum-motoryzacji.com.pl/podstrony/hist_muzeum_ang.html). *www.muzeum-motoryzacji.com.pl*. Archived from the original (http://www.muzeum-motoryzacji.com.pl/podstrony/hist_muzeum_ang.html) on 27 January 2009. . Retrieved 30 January 2009.
[107] (English) "Museum history" (http://muzeumerotyki.com/index.php?lang=en). *muzeumerotyki.com*. . Retrieved 27 August 2011. "The Erotic Museum in Warsaw is a unique place in Poland and one of a few in Europe where you can see pieces of art hidden bashfully from visitors in order museums. (...) Over 2000 exhibits displayed in the Erotic Museum in Warsaw let you explore erotic fantasies of artists from all over the world."
[108] (English) Chris Dziadul. "A decade of progress" (http://www.broadbandtvnews.com/2007/10/05/a-decade-of-progress/). *www.broadbandtvnews.com*. . Retrieved 14 February 2009.
[109] (English) "Press release" (http://www.instytut.com.pl/IMM/o_firmie/Press_release_media_August2008.pdf). *www.instytut.com.pl*. 6 October 2008. . Retrieved 14 February 2009.
[110] (English) "Poland film production guide 2008" (http://www.pisf.pl/pliki/47/ed/47ed315731f90c9/pg2008_i.pdf). *www.pisf.pl*. . Retrieved 14 February 2009.
[111] (English) "The Pianist" (http://web.archive.org/web/20080822051023/http://www.thepianistmovie.com/index2.html). *www.thepianistmovie.com*. Archived from the original (http://www.thepianistmovie.com/index2.html) on 22 August 2008. . Retrieved 14 February 2009.
[112] (English) "Statistical Yearbook of the Republic of Poland 2008" (http://www.stat.gov.pl/cps/rde/xbcr/gus/PUBL_as_statitical_yearbook_of_the_rep_of_poland_2008.pdf). *www.stat.gov.pl*. . Retrieved 30 January 2009.
[113] (Polish) "Studia w liczbach: Warszawa bije Kraków" (http://miasta.gazeta.pl/krakow/1,37650,5009717.html). *miasta.gazeta.pl*. 10 March 2008. . Retrieved 30 January 2009.
[114] (English) "University of Warsaw" (http://www.uw.edu.pl/en/page.php/about_uw/rese.html). *www.uw.edu.pl*. . Retrieved 30 January 2009.
[115] (English) "Warsaw University of Technology (WUT)" (http://www.onelab.eu/index.php/about/management/steering-committee/122-warsaw-university-of-technology-wut.html). *www.onelab.eu*. . Retrieved 30 January 2009. *With over 30,000 students served by over 2,000 professors and instructors, WUT is the largest and the highest-ranking engineering university in Poland.*

[116] (English) "The Fryderyk Chopin University of Music" (http://www.infochopin.pl/en/miejsca.php/99/). *www.infochopin.pl.* . Retrieved 30 January 2009.

[117] (English) "Warsaw School of Economics – Overview" (http://www.sgh.waw.pl/en/ogolne-en/). *www.sgh.waw.pl.* . Retrieved 30 January 2009.

[118] (English) "Warsaw University of Life Sciences" (http://www.sggw.pl/2009/10/12/warsaw-university-of-life-sciences/?lang=en). *www.sggw.pl.* . Retrieved 30 January 2009. *Warsaw University of Life Sciences – SGGW (WULS – SGGW) is the oldest agricultural academic school in Poland, its history dates back to 1816.*

[119] (Polish) "Historia zbiorów" (http://www.bn.org.pl/index.php?id=4). *www.bn.org.pl.* . Retrieved 30 January 2009.

[120] (English) Maria Witt (15 September and 15 October 2005). "The Zaluski Collection in Warsaw" (http://www.fyifrance.com/f102005c.htm). *The Strange Life of One of the Greatest European Libraries of the Eighteenth Century.* FYI France. . Retrieved 17 February 2008.

[121] (English) S.D. Chrostowska. "Polish Literary Criticism Circa 1772: A Genre Perspective" (http://www.utoronto.ca/tsq/14/chrostowska14.shtml). *utoronto.ca.* . Retrieved 17 February 2008.

[122] (Polish) "Historia" (http://www.buw.uw.edu.pl/index.php?option=com_content&task=blogcategory&id=32&Itemid=76). *www.buw.uw.edu.pl.* . Retrieved 30 January 2009.

[123] (Polish) "Zbiory główne" (http://www.buw.uw.edu.pl/index.php?option=com_content&task=blogcategory&id=56&Itemid=121). *www.buw.uw.edu.pl.* . Retrieved 30 January 2009.

[124] (English) "Library building" (http://www.buw.uw.edu.pl/en/index.php?option=com_content&task=view&id=285&Itemid=91). *www.buw.uw.edu.pl.* . Retrieved 30 January 2009.

[125] (English) "Garden" (http://www.buw.uw.edu.pl/en/index.php?option=com_content&task=view&id=286&Itemid=91). *www.buw.uw.edu.pl.* . Retrieved 30 January 2009.

[126] "The most expensive and richest cities in the world" (http://www.citymayors.com/economics/expensive_cities2.html). *UBS.* 18 August 2011. . Retrieved 1 April 2012.

[127] "The World According to GaWC 2008" (http://www.lboro.ac.uk/gawc/world2008t.html). . Retrieved 6 September 2009.

[128] "New MasterCard Research Ranks 65 Cities in Emerging Markets Poised to Drive Long-Term Global Economic Growth | MasterCard" (http://www.mastercard.com/us/company/en/newsroom/pr_new_mastercard_research_ranks_65_Cities_in_emerging_markets.html). Mastercard.com. 22 October 2008. . Retrieved 7 July 2009.

[129] (Polish) "Podmioty gospodarki narodowej" (http://www.stat.gov.pl/cps/rde/xbcr/warsz/ASSETS_podmioty_war_2006.pdf). *www.stat.gov.pl.* 15 February 2007. . Retrieved 28 July 2008.

[130] (English) "Warsaw City Report – March 2007" (http://www.pbwfund.com/_php/get.php/Warsaw city report - March 2007.pdf?id=51&content=application/pdf). *www.pbwfund.com.* . Retrieved 28 July 2008.

[131] (English) "GDP Per Capita" (http://www.stat.gov.pl/cps/rde/xbcr/warsz/ASSETS_09_wwa_18_02.pdf). *www.stat.gov.pl.* . Retrieved 27 October 2010.

[132] (English) "Agriculture and industry" (http://www.pmrconsulting.com/Countries/Poland.html). *www.pmrconsulting.com.* . Retrieved 29 April 2009.

[133] (English) "Big Chance for the Capital" (http://www.warsawvoice.pl/view/18028). *Warsaw – CEE Financial Hub Conference.* www.warsawvoice.pl. 11 June 2008. . Retrieved 28 July 2008.

[134] (English) "History" (http://www.gpw.pl/historia_en). *www.gpw.pl.* . Retrieved 24 May 2012.

[135] "Giełda Papierów Wartościowych w Warszawie" (http://www.gpw.pl/gpw.asp?cel=e_informacje&k=1&i=/periodical_statistic/opis_statistic&sky=1&nagnaz=Information and statistics). Gpw.pl. . Retrieved 22 January 2010.

[136] (English) "Tourism" (http://www.poland.gov.pl/Poland,and,Poles,545.html). *www.poland.gov.pl.* . Retrieved 28 July 2008.

[137] (English) "Industry" (http://www.e-warsaw.pl/ekonomia/przemysl.htm). *www.e-warsaw.pl.* . Retrieved 28 July 2008.

[138] (English) Jerzy J. Parysek. "The socio-economic and spatial transformation of Polish cities after 1989" (http://www.ff.uni-lj.si/oddelki/geo/Publikacije/Dela/files/Dela_21/012 parysek.pdf). *www.ff.uni-lj.si.* . Retrieved 28 July 2008.

[139] http://whc.unesco.org/en/list

[140] http://whc.unesco.org/en/list/30

[141] http://whc.unesco.org/en/list/?search=&search_by_country=&type=&media=&region=&order=region

[142] (English) "Palace" (http://www.wilanow-palac.art.pl/index.php?id=343&menuid=136). *wilanow-palac.art.pl.* . Retrieved 21 February 2008.

[143] (English) Polish Academy of Sciences (1985). *Fragmenta faunistica* (http://books.google.com/?id=VeQaAAAAMAAJ&q=The+Saxon+Garden+(Ogr.+Saski).+The+oldest+public+park+in+Warsaw,+founded+in+1713.&dq=The+Saxon+Garden+(Ogr.+Saski).+The+oldest+public+park+in+Warsaw,+founded+in+1713.). Instytut zoologii, Polska Akademia Nauk. p. 317. . *The Saxon Garden (Ogr. Saski). The oldest public park in Warsaw, founded in 1713.*

[144] (Polish) "Historia" (http://www.lazienki-krolewskie.pl/). *www.lazienki-krolewskie.pl.* . Retrieved 31 January 2009.

[145] "Short and long history of the Powiązki Cemetery" (http://ceo.rsi.pl/dokument.php?dzial=2143&id=48006&PHPSESSID=f6c400ad7970068b81fd2708234f8f93) (in Polish). . Retrieved 11 February 2008.

[146] (English) "Warsaw Judaica" (http://web.archive.org/web/20070805045632/http://www.um.warszawa.pl/v_syrenka/perelki/index_en.php?mi_id=270&dz_id=18). *www.um.warszawa.pl.* Archived from the original (http://www.um.warszawa.pl/v_syrenka/perelki/index_en.php?mi_id=270&dz_id=18) on 5 August 2007. . Retrieved 26 January 2010.

[147] (English) "Heroic City" (http://www.um.warszawa.pl/v_syrenka/perelki/index_en.php?mi_id=174&dz_id=17). *www.um.warszawa.pl*. . Retrieved 26 January 2010.
[148] (English) James Ramsay Montagu Butler, Norman Henry Gibbs, J. M. A. Gwyer, John Patrick William Ehrman, Michael Eliot Howard (1976). "History of the Second World War; United Kingdom military series 5". In James Ramsay Montagu Butler. *Grand strategy*. H. M. Stationery Off. p. 369.
[149] (English) "Church of the Holy Cross" (http://web.archive.org/web/20060218162926/http://um.warszawa.pl/v_syrenka/perelki/index_en.php?mi_id=112&dz_id=16). *eGuide / Treasures of Warsaw on-line*. Archived from the original (http://um.warszawa.pl/v_syrenka/perelki/index_en.php?mi_id=112&dz_id=16) on 18 February 2006. . Retrieved 23 February 2009.
[150] (English) "Frédéric Chopin Monument" (http://um.warszawa.pl/v_syrenka/perelki/index_en.php?mi_id=165&dz_id=16). *eGuide / Treasures of Warsaw on-line*. . Retrieved 23 February 2009.
[151] (English) "Polish Girlhood (1867–1891)" (http://www.aip.org/history/curie/polgirl2.htm). *www.aip.org*. American Institute of Physics. . Retrieved 25 February 2009.
[152] (English) "The Radium Institute (1919–1934)" (http://www.aip.org/history/curie/radinst1.htm). *www.aip.org*. American Institute of Physics. . Retrieved 25 February 2009.
[153] "The Mermaid" (http://www.ucl.ac.uk/atlas/polish/mywarsaw/warsaw10.html). . Retrieved 11 February 2008.
[154] "Warsaw Mermaid's Statue" (http://web.archive.org/web/20081207092210/http://www.um.warszawa.pl/v_syrenka/perelki/index_en.php?mi_id=47&dz_id=2). Archived from the original (http://www.um.warszawa.pl/v_syrenka/perelki/index_en.php?mi_id=47&dz_id=2) on 7 December 2008. . Retrieved 10 July 2008.
[155] (English) "History of Warsaw's Coat of Arms" (http://www.e-warsaw.pl/miasto/herb.htm). *www.e-warsaw.pl*. . Retrieved 10 July 2008.
[156] (English) Ewa Bratosiewicz. "Other symbols of Warsaw" (http://www.warsaw-guide.invito.pl/index.php?str=x41). *www.warsaw-guide.invito.pl*. . Retrieved 10 July 2008.
[157] "Warsaw Mermaid – Syrena" (http://biega.com/syrena.html). . Retrieved 10 July 2008.
[158] (English) "History of Warsaw's Coat of Arms" (http://www.e-warsaw.pl/miasto/herb-1.htm). *www.e-warsaw.pl*. . Retrieved 10 July 2008.
[159] (English) "The Maid of Warsaw" (http://www.qohmuseum.org.uk/maid.htm). *The Queen's Own Hussars Museum*. . Retrieved 10 July 2008.
[160] (English) "RAF Odiham" (http://www2.army.mod.uk/linkedfiles/soldierwelfare/supportagencies/aws/communityguides/swf_sa_aws_cg_w/raf_odiham_2008/raf_odiham_section_1.pdf). *www2.army.mod.uk*. p. 16. . Retrieved 10 July 2008.
[161] (English) "Marie Curie – The Nobel Prize in Physics 1903" (http://nobelprize.org/nobel_prizes/physics/laureates/1903/marie-curie-bio.html). *nobelprize.org*. . Retrieved 10 July 2008.
[162] (English) Joanna Ławrynowicz. "Frederick Francois Chopin, the most eminent Polish composer" (http://www.infochopin.pl/en/artykuly.php/1/). *www.infochopin.pl*. . Retrieved 10 July 2008.
[163] (English) *Women artists in the 20th and 21st century* (http://books.google.com/?id=ZSvSfCmzo2wC&printsec=frontcover). Taschen. 2001. p. 576. ISBN 3-8228-5854-4. .
[164] (Polish) "Tamara Łempicka" (http://www.marchand.pl/artysta.php?id=134&biografia=f&l=pl). *www.marchand.pl*. . Retrieved 22 January 2009.
[165] "Moshe Vilensky" (http://www.jewishvirtuallibrary.org/jsource/biography/Vilensky.html). Jewishvirtuallibrary.org. . Retrieved 31 July 2011.
[166] (Polish) "The 5th Festival of Jewish Culture 'Singer's Warsaw'" (http://www.jewish-theatre.com/visitor/article_display.aspx?articleID=2973). *www.jewish-theatre.com*. . Retrieved 4 March 2009.
[167] (English) Richard Burgin, Issac Bashevis Singer (1978). *Issac bashevis Singer Talks... About Everything*. The New York Times Magazine. p. 46. in: David Neal Miller, Isaac Bashevis Singer (1986). *Recovering the canon: essays on Isaac Bashevis Singer* (http://books.google.com/?id=O6hNoUXlCXwC&pg=PA40&dq=Warsaw+has+just+now+been+destroyed.+No+one+will+ever+see+the+Warsaw+I+knew.+Let+me+just+write+about+it.+Let+this+Warsaw+not+disappear+forever#v=onepage&q=Warsaw has just now been destroyed.No one will ever see the Warsaw I knew.Let me just write about it.Let this Warsaw not not disappear forever&f=false). BRILL. p. 40. ISBN 90-04-07681-6. .
[168] "Miasta partnerskie Warszawy" (http://www.um.warszawa.pl/node/2920?page=0,0). *um.warszawa.pl*. Biuro Promocji Miasta. 4 May 2005. . Retrieved 29 August 2008.
[169] "Berlin's international city relations" (http://www.berlin.de/rbmskzl/staedteverbindungen/warschau.en.html). Berlin Mayor's Office. . Retrieved 1 July 2009. *Berlin and Warsaw's agreement on friendship and cooperation and a corresponding supporting program was signed in Berlin on 12 August 1991.*
[170] (Spanish) "Listado de ciudades hermanas" (http://www.buenosaires.gov.ar/areas/internacionales/hermanamientos.pdf). *www.buenosaires.gov.ar*. Gobierno de la Ciudad de Buenos Aires. . Retrieved 15 October 2011. "1990. Praga. 1992. Rótterdam. 1990. Varsovia."
[171] "Twin Towns" (http://www.amazingdusseldorf.com/community-local/people/twin-towns.html). www.amazingdusseldorf.com. . Retrieved 29 October 2009.
[172] "Sister Cities of Istanbul" (http://www.greatistanbul.com/sister_cities.htm). . Retrieved 8 September 2007.

[173] Erdem, Selim Efe (3 November 2003). "İstanbul'a 49 kardeş" (http://www.radikal.com.tr/haber.php?haberno=94185) (in Turkish). Radikal. . "49 sister cities in 2003"
[174] Madrid city council webpage "Mapa Mundi de las ciudades hermanadas" (http://www.munimadrid.es/portal/site/munimadrid/menuitem.dbd5147a4ba1b0aa7d245f019fc08a0c/?vgnextoid=4e84399a03003110VgnVCM2000000c205a0aRCRD&vgnextchannel=4e98823d3a37a010VgnVCM100000d90ca8c0RCRD&vgnextfmt=especial1&idContenido=508c7aefd9b5b010VgnVCM100000d90ca8c0RCRD). Ayuntamiento de Madrid. Madrid city council webpage.
[175] "Partners – Oslo kommune" (http://web.archive.org/web/20090102135652/http://www.oslo.kommune.no/the_city_of_oslo/international_cooperation/). www.oslo.kommune.no. Archived from the original (http://www.oslo.kommune.no/the_city_of_oslo/international_cooperation/) on 2 January 2009. .
[176] "Twin cities of Riga" (http://www.riga.lv/EN/Channels/Riga_Municipality/Twin_cities_of_Riga/default.htm). Riga City Council. . Retrieved 27 July 2009.
[177] "Saint Petersburg in figures – International and Interregional Ties" (http://eng.gov.spb.ru/figures/ities). Saint Petersburg City Government. . Retrieved 23 March 2008.
[178] "Online Directory: California, USA" (http://web.archive.org/web/20080116164532/http://www.sister-cities.org/icrc/directory/usa/CA). Sister Cities International. Archived from the original (http://www.sister-cities.org/icrc/directory/usa/CA) on 16 January 2008. . Retrieved 22 April 2009.
[179] Seul Metropolitan Government. "International Cooperation: Sister Cities" (http://english.seoul.go.kr/gover/cooper/coo_02sis.html). .
[180] Sister city list (http://www.edunet.taipei.gov.tw/attach/The 45 Sister Cities list.doc) (.DOC)
[181] "Tel Aviv sister cities" (http://web.archive.org/web/20090214183503/http://www.tel-aviv.gov.il/Hebrew/Cityhall/TwinCities/Index.asp) (in Hebrew). Tel Aviv-Yafo Municipality. Archived from the original (http://www.tel-aviv.gov.il/Hebrew/Cityhall/TwinCities/Index.asp) on 14 February 2009. . Retrieved 14 July 2009.
[182] (Polish) "Miasta partnerskie Warszawy" (http://web.archive.org/web/20100704163612/http://um.warszawa.pl/v_syrenka/new/index.php?dzial=aktualnosci&ak_id=3284&kat=11). *um.warszawa.pl*. Biuro Promocji Miasta. 4 May 2005. Archived from the original (http://um.warszawa.pl/v_syrenka/new/index.php?dzial=aktualnosci&ak_id=3284&kat=11) on 4 July 2010. . Retrieved 29 August 2008.
[183] http://books.google.com/?id=E4cM2Hf8KYsC&printsec=frontcover#v=onepage&q&f=false
[184] http://books.google.com/books?id=xXbqAQAACAAJ

External links

- Official web page of Warsaw (http://www.e-warsaw.pl/index.php) includes 360° panoramas of the UNESCO listed (http://www.e-warsaw.pl/miasto/unesco.htm) region
- District Police Headquarters – Warsaw II (http://mokotow.policja.waw.pl/?page=Structure&id=131) (part of Warsaw Metropolitan Police)
- Warsaw Tourist Guide (http://www.warsawguide.com/) – information for travellers
- Warsaw message board (http://www.polished.pl/forum) – information and advice for foreigners living in Warsaw
- Portal Warszawa City (http://www.warszawacity.pl/) – Polish

Coat of arms

Escutcheon

Field

Supporter

Supporter

Motto (Scotland)

Crest

Torse

Mantling

Helm

Coronet

Compartment

Order

Ordinaries

Common

charges

Motto

Achievement elements

A **coat of arms** is a unique heraldic design on a shield or escutcheon or on a surcoat or tabard used to cover and protect armour and to identify the wearer. Thus the term is often stated as "coat-armour", because it was anciently displayed on the front of a coat of cloth. The coat of arms on an escutcheon forms the central element of the full heraldic achievement which consists of shield, supporters, crest and motto. The design is a symbol unique to an individual person, and to his family, corporation, or state. Such displays are commonly called *armorial bearings*, *armorial devices*, *heraldic devices*, or simply *armorials* or *arms*.

Historically, armorial bearings were first used by feudal lords and knights in the mid-12th century on battlefields as a way to identify allied from enemy soldiers. As the uses for heraldic designs expanded, other social classes who never would march in battle began to assume arms for themselves. Initially, those closest to the lords and knights adopted arms, such as persons employed as squires that would be in common contact with the armorial devices. Then priests and other ecclesiastical dignities adopted coats of arms, usually to be used as seals and other such insignia, and then towns and cities to likewise seal and authenticate documents. Eventually by the mid-13th century, peasants, commoners and burghers were adopting heraldic devices. The widespread assumption of arms led some states to regulate heraldry within their borders. However, in most of continental Europe, citizens freely adopted armorial bearings.

Despite no widespread regulation, and even with a lack in many cases of national-level regulation, heraldry has remained rather consistent across Europe, where traditions alone have governed the design and use of arms. Unlike seals and other general emblems, heraldic achievements have a formal description called a blazon, expressed in a jargon that allows for consistency in heraldic depictions.

In the 21st century, coats of arms are still in use by a variety of institutions and individuals; for example, universities have guidelines on how their coats of arms may be used, and protect their use as trademarks.[1] [2] [3] Many societies exist that also aid in the design and registration of personal arms, and some nations, like England and Scotland, still maintain to this day the mediæval authorities that grant and regulate arms.

Traditions and usage

In the heraldic traditions of England and Scotland an individual, rather than a family, had a coat of arms. In those traditions coats of arms are legal property transmitted from father to son; wives and daughters could also bear arms modified to indicate their relation to the current holder of the arms. Undifferenced arms are used only by one person at any given time. Other descendants of the original bearer could bear the ancestral arms only with some difference: usually a color change or the addition of a distinguishing charge. One such charge is the label, which in British usage (outside the Royal Family) is now always the mark of an heir apparent or (in Scotland) an heir presumptive.

The German *Hyghalmen Roll*, ca. late 15th century, illustrates the German practice of thematic repetition from the arms in the crest

Because of their importance in identification, particularly in seals on legal documents, the use of arms was strictly regulated; few countries continue in this today. This has been carried out by heralds and the study of coats of arms is therefore called "heraldry". Some other traditions (e.g., Polish heraldry) are less restrictive — allowing, for example, all members of a dynastic house or family to use the same arms, although one or more elements may be reserved to the head of the house.

In time, the use of arms spread from military entities to educational institutes, and other establishments.[2] According to a design institute article, "The modern logo and corporate livery have evolved from the battle standard and military uniform of medieval times".[4]

In his book, *The Visual Culture of Violence in the Late Middle Ages*, Valentin Groebner argues that the images composed on coats of arms are in many cases designed to convey a feeling of power and strength, often in military terms.[5] The author Helen Stuart argues that some coats of arms were a form of corporate logo.[6] Museums on medieval armory also point out that as emblems they may be viewed as precursors to the corporate logos of modern society, used for group identity formation.[7] [8]

When knights were so encased in armour that no means of identifying them was left, the practice was introduced of painting their insignia of honour on their shield as an easy method of distinguishing them. Originally these were granted only to individuals, but were afterward made hereditary in England by King Richard I, during his crusade to Palestine.

European tradition

British heraldry

In Scotland, the Lord Lyon King of Arms has criminal jurisdiction to enforce the laws of arms. In England, Northern Ireland and Wales the use of arms is a matter of civil law and regulated from the College of Arms and the Court of Chivalry.

In reference to a dispute over the exercise of authority over the Officers of Arms in England, Arthur Annesley, 1st Earl of Anglesey, Lord Privy Seal, declared on 16 June 1673 that the powers of the Earl Marshal were "to order, judge, and determine all matters touching arms, ensigns of nobility, honour, and chivalry; to make laws, ordinances, and statutes for the good government of the Officers of Arms; to nominate Officers to fill vacancies in the College of Arms; to punish and correct Officers of Arms for misbehaviour in the execution of their places". It was further declared that no patents of arms or any ensigns of nobility should be granted and no augmentation, alteration, or addition should be made to arms without the consent of the Earl Marshal.

The coat of arms of the University of Plymouth, which were granted in 2008 by the College of Arms, shows the heater shield and style popular in modern British heraldry.[9]

Irish heraldry

In Ireland, since 1552, the usage and granting of coats of arms has been strictly regulated by the Government of Ireland, through the Office of the Chief Herald of Ireland.

German and Scandinavian heraldry

The heraldic tradition and style of modern and historic Germany and the Holy Roman Empire — including national and civic arms, noble and burgher arms, ecclesiastical heraldry, heraldic displays and heraldic descriptions — stand in contrast to Gallo-British, Latin and Eastern heraldry, and strongly influenced the styles and customs of heraldry in the Nordic countries, which developed comparatively late.[10]

In the Nordic countries, provinces, regions, cities and municipalities have a coat of arms. These are posted at the borders and on buildings containing official offices, as well as used in official documents and on the uniforms of municipal officers. Arms may also be used on souvernirs or other effects, given that an application has been granted by the municipal council.

Coat of arms of the city of Vaasa, showing the shield with the Royal House of Wasa emblem, a crown and a Cross of Liberty pendant.

Other European countries

At a national level, "coats of arms" were generally retained by European states with constitutional continuity of more than a few centuries, including constitutional monarchies like Denmark as well as old republics like San Marino and Switzerland.

In Ireland the usage and granting of coats of arms is strictly regulated by the Government of Ireland through the Office of the Chief Herald of Ireland.

In Italy the use of coats of arms was only loosely regulated by the states existing before the unification of 1860. Since the Consulta Araldica, the college of arms of the Kingdom of Italy, was abolished in 1948, personal coats of arms and titles of nobility, though not outlawed, are not recognised.

Among the states ruled by communist regimes, coats of arms resembling the Soviet design were adopted in all the Warsaw Pact states except Czechoslovakia and Poland.

Since 1989, some of the ex-Communist states, such as Romania, have resumed their former arms, often with only the symbols of monarchy removed.

Practice in the Americas

Canada

The Queen of Canada has delegated her prerogative to grant armorial bearings to the Governor General of Canada. Canada has its own Chief Herald and Herald Chancellor. The Canadian Heraldic Authority is situated at Rideau Hall.[11] [12]

United States

The Great Seal of the United States uses on the obverse as its central motif an heraldic achievement described as being the arms of the nation.[13] The seal, and the armorial bearings, were adopted by the Continental Congress on June 20, 1782, and is a shield divided palewise into thirteen pieces, with a blue chief, which is displayed upon the breast of an American bald eagle. The crest is thirteen stars breaking through a glory and clouds, displayed with no helm, torse or mantling (unlike most European precedents).

The Great Seal of the United States, which displays as its central design the heraldic device of the nation.

Only a few of the American states have adopted a coat of arms, which is usually designed as part of the respective state's seal. Vermont has both a state seal and a state coat of arms that are independent of one another (though both contain a pine tree, a cow and sheaves of grain); the seal is used to authenticate documents, whilst the heraldic device represents the state itself.

Ecclesiastic practice

Catholic Church

Vatican City State and the Holy See each have their own coat of arms. As the Papacy is not hereditary, its occupants display their personal arms combined with those of their office.

Some Popes came from armigerous (noble) families; others adopted coats of arms during their career in the church. The latter typically allude to their ideal of life, or to specific Pontifical programmes.[14] A well known and widely displayed example in recent times was Pope John Paul II's coat of arms. His selection of a large letter M (for Mary) on his coat of arms was intended to express the message of his strong Marian devotion.[15]

Roman Catholic Dioceses also are assigned a coat of arms. A Basilica, or papal church also gets a coat of arms, which is usually displayed on the building. These may be used in countries which otherwise do not use heraldic devices.

However in countries like Scotland with a strong statutory heraldic authority arms will need to be officially granted and recorded.

Coat of arms of Pope John Paul II, display the Papal tiara and crossed keys of the pontifical office.

Arab World

Egyptian coats of arms showing common Near and Middle Eastern motifs, namely the crescent and stars which are symbols of the region's predominant religion, Islam, and the eagle which is widely believed to have been an emblem used by Saladin.

With the formation of the modern nation states of the Arab World in the second half of the 20th century, European traditions of heraldry were partially adopted for state emblems.

These emblems often involve the star and crescent symbol taken from the Ottoman flag. Another commonly seen charge is the eagle, which is a symbol attributed to Saladin, and the hawk of the Qureish.

Similar traditions

Japan

The Japanese equivalents, called *kamon* (often abbreviated "mon"), are family badges which often date back to the 7th century, and are still actively used in Japan today. The Japanese tradition is independent of the European, and thus very different in style; but as in Europe many abstract and floral elements are used.

Yet, even these simple designs often express an origin. An example in recent use is the logo of Mitsubishi corporation which started as a shipping and maritime enterprise and whose emblem is based on a water chestnut derived from its maritime history with a military naval influence. The word *mitsu* means the number 3 and the word *hishi* meaning "water chestnut" (pronounced *bishi* in some combinations; see rendaku) originated from the emblem of the warrior Tosa Clan. The battleships of the Tosa Clan had been used in the late 19th century in the First Sino-Japanese War to reach Korea[16] and their name was given to a modern battleship. The Tosa water chestnut leaf mon was then drawn as a rhombus or diamond shape in the Mitsubishi logo.[17] [18]

Flags and banners

Flags are used to identify ships (where they are called ensigns), embassies and such, and they use the same colors and designs found in heraldry, but they are not usually considered to be heraldic. A country may have both a national flag and a national coat of arms, and the two may not look alike at all. For example, the flag of Scotland (St Andrew's Cross) has a white saltire on a blue field, but the royal arms of Scotland has a red lion within a double tressure on a gold (or) field.

See also

- List of coats of arms
- Baron and feme
- Siebmachers Wappenbuch (Coats of arms from Germany, Switzerland, Austria, etc.)

Notes

[1] Glasgow University - The Coat of Arms (http://www.universitystory.gla.ac.uk/coat-of-arms/)

[2] Educational Institute Coat of arms (http://www.emeraldinsight.com/Insight/ViewContentServlet?Filename=Published/EmeraldFullTextArticle/Articles/1680100103.html)

[3] Policy on use of the Workmark and Insignia of McGill University (http://www.mcgill.ca/files/secretariat/WordmarkandInsigniaofMcGillUniversity-Policyontheuseof.pdf)

[4] Understanding Corporate Identity (http://www.dia.org.au/content.cfm?news=30&id=102)

[5] Valentin Groebner, 2004, *The Visual Culture of Violence in the Late Middle Ages* ISBN 978-1-890951-37-5

[6] *Employee Identification with the Corporate Identity* International Studies of Management and Organization, Volume 32, Number 3, 2002

[7] Group Identity Formation in the German Renaissance (http://www.phil-hum-ren.uni-muenchen.de/GermLat/Acta/Bernstein.htm)

[8] European Coat of Arms Museum (http://www.wien.gv.at/english/culture/museums/coat-of-arms.htm)

[9] College of Arms (http://www.college-of-arms.gov.uk/Newsletter/018.htm)

[10] Volborth, Carl-Alexander von (1981). *Heraldry: Customs, Rules and Styles*. Poole, England: Blandford Press. ISBN 0-7137-0940-5. ISBN 0-7137-0940-5 p.129.

[11] "The History of Heraldry in Canada" (http://www.heraldry.ca/top_en/top_historyHer.htm). Royal Heraldry Society of Canada. 28 April 2004. . Retrieved 2008-08-21.

[12] "The Canadian Heraldic Authority" (http://www.gg.ca/heraldry/cha/index_e.asp). Canadian Heraldic Authority. 2005-09-27; updated 2006-06-14. . Retrieved 2008-09-02.

[13] 2004 Seal Broch (http://www.state.gov/documents/organization/27807.pdf)

[14] Coat of Arms of His Holiness Benedict XVI (http://www.vatican.va/holy_father/benedict_xvi/elezione/stemma-benedict-xvi_en.html)

[15] Vatican press office (http://www.vatican.va/news_services/press/documentazione/documents/sp_ss_scv/insigne/sp_ss_scv_stemma-bandiera-sigillo_en.html)

[16] Donald Calman, 1992 *The Nature and Origins of Japanese Imperialism* ISBN 0-415-06710-3

[17] Mitsubishi Mark (http://www.mitsubishi.com/e/group/mark.html)

[18] Mitsubishi History (http://www.mitsubishi.com/e/history/index.html)

References

- Pimbley, Arthur Francis (1908). *Pimbley's dictionary of heraldry*. Pimbley.
- This article incorporates text from a publication now in the public domain: "Pimbley's dictionary of heraldry", Arthur Francis Pimbley (1908)

External links

- College of Arms (http://www.college-of-arms.gov.uk/): Repository of the coats of arms and pedigrees of English, Welsh, Northern Irish and Commonwealth families and their descendants together with, and in principle under the control of, the legal body the Court of Chivalry, both medieval in origin.
- The Court of the Lord Lyon (http://www.lyon-court.com/lordlyon/CCC_FirstPage.jsp): the statutory heraldry office for Scotland
- Free access (http://www.luz-herald.net/free/) to Burke's General Armory (incomplete, 1500 British surnames), Pimbley's Dictionary of Heraldry and Blason des familles d'Europe, Grand Armorial Universel (15,000 European surnames)
- Royal Dutch Library page for the "Wapenboek Beyeren" (http://www.kb.nl/galerie/wapenboek/index.html/) written by Claes Heynenzoon around 1400, containing over 1000 drawings of coats of arms
- General armorial of noble families in the Russian Empire (Gerbovnik) (http://gerbovnik.com)

Canting arms

Canting arms are heraldic bearings that represent the bearer's name in a visual pun or rebus. The term *cant* came into the English language from Anglo-Norman *cant*, meaning song or singing, from Latin *cantāre*, and English cognates include *canticle, chant, accent, incantation* and *recant.*[2]

A common simple example of canting arms: the *castle* representing the Kingdom of *Castile* and the *lion* representing the Kingdom of *León.*[1]

Canting arms – some in the form of rebuses – are quite common in German civic heraldry. They have also been increasingly used in the 20th century among the British royal family. When the visual representation is not straightforward but as complex as a rebus, this is sometimes called a *rebus coat of arms*. An in-joke among Society for Creative Anachronism heralds is the pun, "Heralds don't pun; they cant."[3]

Examples of canting arms

A famous example of canting arms are those of the late Queen Elizabeth, the Queen Mother, who was born Elizabeth Bowes-Lyon. Her arms (pictured below) contain in sinister (i.e. on the bearer's left, viewer's right) the bows and blue lions that make up the arms of the Bowes and Lyon families.

Elizabeth Bowes-Lyon: bows and lions

Princess Beatrice of York: *Beatrice* = bee thrice = three bees

Sometimes also called "canting" are municipal coats of arms which interpret the town's name in rebus form

The arms of Berwickshire, Scotland: Bear and Wych Elm

Elmbridge, Surrey (1974): elm tree on bridge. (The toponym is related to bridges but not to elms; the prefix refers to *Emel*, a former name for the river Mole.[4])

Châteaurenard: *Château* = castle; *Renard* = fox

Eberbach (1976): *Eber* = boar; *Bach* = brook (wavy blue fess)

Hensbroek (1817[5]): The coat of arms of the village of Hensbroek in North Holland interprets the toponym as "hen-breeches" (the toponym is unrelated to either "hen" or "breeches", deriving from a personal name *Hein* and the Dutch cognate of "brook", i.e. "Henry's brook".)

Freixo de Espada à Cinta (1926): *Freixo* = ash; *de Espada* = with sword; *à Cinta* = at the waist, in Portuguese

Falkenberg (1971): *Falken* = falcon; *Berg* = hill, in Swedish

Seinäjoki (1951): *seinä* = wall, *joki* = river, in Finnish

Notes

[1] http://www.heraldica.org/topics/canting.htm

[2] "Cant". *The American Heritage Dictionary of the English Language: Fourth Edition*. 2000. ISBN 0-395-82517-2.

[3] Neznanich, Modar. "Heraldry for Those Who Cant" (http://www.modaruniversity.org/Heraldry-for-Those-Who-Cant.pdf). . Retrieved 2 July 2012. Cites 72 historical examples of canting arms, as well as SCA usage.

[4] Room, Adrian (1988). *Dictionary Of Place Names In The British Isles*. Bloomsbury. p. 128.

[5] Weeks, Andrew. "Obdam (The Netherlands)" (http://www.crwflags.com/fotw/flags/nl-nh-od.html). *Flags of the World*. CRW Flags. . Retrieved 16 September 2011.

References

- "Meaning of Arms" (http://www.heraldica.org/topics/meaning.htm). *Heraldica.org*. 2001-06-20.

Łódź Voivodeship

Łódź Voivodeship
Województwo łódzkie

— Voivodeship —

Flag

Coat of arms

Location within Poland

Division into counties

Coordinates (Łódź): 51°47′N 19°28′E	
Country	Poland
Capital	Łódź
Counties	
Area	
• Total	18219 km^2 (**unknown operator: u'strong'** sq mi)
Population (2006)	
• Total	2571534
• Density	**unknown operator: u'strong'**/km^2 (**unknown operator: u'strong'**/sq mi)
• Urban	1662937
• Rural	908597
Car plates	E
Website	http://www.lodzkie.pl
* further divided into 177 gminas	

Łódź Voivodeship (also known as **Łódź Province**,[1] or by its Polish name of *województwo łódzkie* [vɔjɛˈvut͡stfɔ ˈwut͡skʲɛ] or simply *Łódzkie*) is a province (voivodeship) in central Poland. It was created on 1 January 1999 out of the former Łódź Voivodeship (1975–1999) and the Sieradz, Piotrków Trybunalski and Skierniewice Voivodeships and part of Płock Voivodeship, pursuant to the Polish local government reforms adopted in 1998. The province is named after its capital and largest city, Łódź, pronounced [wut͡ɕ].

Łódź Voivodeship is bordered by six other voivodeships: Masovian to the north and east, Świętokrzyskie to the south-east, Silesian to the south, Opole to the south-west, Greater Poland to the west, and Kuyavian-Pomeranian for a short stretch to the north. Its territory belongs to three historical provinces of Poland – Masovia (in the east), Greater Poland (in the west) and Lesser Poland (in the southeast, around Opoczno).

Cities and towns

The voivodeship contains 44 cities and towns. These are listed below in descending order of population (according to official figures for 2006):[2]

1. Łódź (764,168)
2. Piotrków Trybunalski (79,367)
3. Pabianice (70,445)
4. Tomaszów Mazowiecki (66,705)
5. Bełchatów (62,062)
6. Zgierz (58,313)
7. Radomsko (49,152)
8. Skierniewice (48,761)
9. Kutno (47,557)
10. Zduńska Wola (44,370)
11. Sieradz (44,045)
12. Łowicz (30,204)
13. Wieluń (24,347)
14. Opoczno (22,708)
15. Ozorków (20,571)

1. Aleksandrów Łódzki (20,512)
2. Łask (18,684)
3. Rawa Mazowiecka (17,643)
4. Konstantynów Łódzki (17,564)
5. Łęczyca (15,423)
6. Głowno (15,167)
7. Koluszki (13,407)
8. Brzeziny (12,373)
9. Żychlin (8,880)
10. Wieruszów (8,759)
11. Zelów (8,173)
12. Poddębice (7,875)
13. Tuszyn (7,178)
14. Pajęczno (6,674)
15. Sulejów (6,387)

1. Działoszyn (6,276)
2. Krośniewice (4,647)
3. Drzewica (3,945)
4. Przedbórz (3,758)
5. Stryków (3,566)
6. Złoczew (3,403)
7. Warta (3,388)
8. Rzgów (3,338)
9. Biała Rawska (3,182)
10. Uniejów (2,916)
11. Kamieńsk (2,858)
12. Wolbórz (2,381)
13. Błaszki (2,179)
14. Szadek (2,007)

Administrative division

Łódź Voivodeship is divided into 24 counties (powiats): 3 city counties and 21 land counties. These are further divided into 177 gminas.

The counties are listed in the following table (ordering within categories is by decreasing population).

English and Polish names	Area (km²)	Population (2006)	Seat	Other towns	Total gminas
City counties					
Łódź	293	764,168			1
Piotrków Trybunalski	67	79,367			1
Skierniewice	33	48,761			1
Land counties					
Zgierz County powiat zgierski	854	160,805	Zgierz	Ozorków, Aleksandrów Łódzki, Głowno, Stryków	9
Sieradz County powiat sieradzki	1,491	121,013	Sieradz	Złoczew, Warta, Błaszki	11
Tomaszów Mazowiecki County powiat tomaszowski	1,026	120,973	Tomaszów Mazowiecki		11
Pabianice County powiat pabianicki	491	119,008	Pabianice	Konstantynów Łódzki	7
Radomsko County powiat radomszczański	1,443	118,856	Radomsko	Przedbórz, Kamieńsk	14
Bełchatów County powiat bełchatowski	969	112,640	Bełchatów	Zelów	8
Kutno County powiat kutnowski	886	104,124	Kutno	Żychlin, Krośniewice	11

Piotrków County powiat piotrkowski	1,429	90,227	Piotrków Trybunalski *	Sulejów, Wolbórz	11
Łowicz County powiat łowicki	987	82,338	Łowicz		10
Opoczno County powiat opoczyński	1,039	78,659	Opoczno	Drzewica	8
Wieluń County powiat wieluński	928	78,260	Wieluń		10
Zduńska Wola County powiat zduńskowolski	369	67,704	Zduńska Wola	Szadek	4
Łódź East County powiat łódzki wschodni	499	64,574	Łódź *	Koluszki, Tuszyn, Rzgów	6
Łęczyca County powiat łęczycki	774	53,435	Łęczyca		8
Pajęczno County powiat pajęczański	804	53,395	Pajęczno	Działoszyn	8
Łask County powiat łaski	617	50,874	Łask		5
Rawa County powiat rawski	647	49,443	Rawa Mazowiecka	Biała Rawska	6
Wieruszów County powiat wieruszowski	576	42,336	Wieruszów		7
Poddębice County powiat poddębicki	881	42,195	Poddębice	Uniejów	6
Skierniewice County powiat skierniewicki	756	37,779	Skierniewice *		9
Brzeziny County powiat brzeziński	359	30,600	Brzeziny		5
* seat not part of the county					

Protected areas

Protected areas in Łódź Voivodeship include seven Landscape Parks, as listed below.

- Bolimów Landscape Park (partly in Masovian Voivodeship)
- Łódź Hills Landscape Park
- Przedbórz Landscape Park (partly in Świętokrzyskie Voivodeship)
- Spała Landscape Park
- Sulejów Landscape Park
- Warta-Widawka Landscape Park
- Załęcze Landscape Park (partly in Silesian Voivodeship)

History

The capital of the Łódź Voivodeship has always been Łódź, but the area of land which it comprises has changed several times. The first was a unit of administrative division and local government in the Second Polish Republic in the years 1921–1939. In 1938 some western counties were ceded to Greater Poland Voivodeship (see: Territorial changes of Polish Voivodeships on April 1, 1938).

Łódź Voivodeship 1921–1939

After the change, Łódź Voivodeship's area was 20,446 km², and its population (as for 1931) was 2,650,100. It consisted of 15 powiats (counties):

- Brzeziny county,
- Końskie county,
- Kutno county,
- Łask county,
- Łęczyca county,
- Łowicz county,
- city of Łódź county (powiat lodzki grodzki),
- Łódź county,
- Opoczno county,
- Piotrków Trybunalski county,
- Radomsko county,
- Rawa Mazowiecka county,
- Sieradz county,
- Skierniewice county,
- Wieluń county.

The largest cities of the Voivodeship were (population according to the 1931 census):

- Łódź (pop. 604,600),
- Piotrków Trybunalski (pop. 51,300),
- Pabianice (pop. 45,700),
- Tomaszów Mazowiecki (pop. 38,000),
- Zgierz (pop. 26,600),
- Kutno (pop. 23,400),
- Radomsko (pop. 23,000).

Source: Maly rocznik statystyczny 1939, Nakladem Glownego Urzedu Statystycznego, Warszawa 1939 (Concise Statistical Year-Book of Poland, Warsaw 1939).

The next incarnation existed from 1945 until 1975 (although the city of Łódź was excluded as a separate City Voivodeship). This Łódź Voivodeship was then broken up, superseded by Łódź (see below), Sieradz, Piotrków Trybunalski, Skierniewice and partly Płock Voivodeships.

Łódź Voivodeship, also known as Łódź Metropolitan Voivodeship (*województwo miejskie łódzkie*), existed from 1975 until 1998, after which it was incorporated into today's Łódź Voivodeship. Until 1990, the mayor of the city of Łódź was also the voivodeship governor.

As of 1995, major cities and towns in Łódź Metropolitan Voivodeship included (with their 1995 populations):

- Łódź (825,600);
- Pabianice (75,700);
- Zgierz (59,100);
- Ozorków (21,900);
- Aleksandrów Łódzki (20,400).

Łódź Voivodeship 1975–1998

Culture and education

The basic activities of the Lodz Region in terms of culture are: monitoring activities of seven regional self-government cultural institutions, i.e.: the Arthur Rubinstein Lodz Philharmonic, Museum of Art in Łódź (having one of the biggest modern art collections in Europe), the Opera House, Stefan Jaracz Theater, the Museum of Archeology and Ethnography, The Józef Piłsudski Regional and Municipal Public Library in Lodz, the Chamber of Culture in Lodz but also: supporting NGO's, protection of monuments, awarding scholarships to young artists and rewards for the prominent artists. What is more, infrastructural projects are being undertaken. Among the most important investments are: the creation of four regional scenes in Stefan Jaracz Theater, opening the new section of the Museum of Art in Lodz - ms² or the reconstruction of medieval settlement in Tum in the vicinity of Łęczyca. The major universities in Łódź Voivodeship are:

- University of Łódź
- Technical University of Łódź
- National Film School in Łódź
- Medical University of Łódź
- Academy of Fine Arts and Design in Łódź
- Jan Kochanowski University in Piotrków Trybunalski

There are also dozens of other schools and academies, but for the last four years the best students in Łódź Voivodeship (according to the prestigious contest "Studencki Nobel") have been studying at the University of Łódź - in 2009 the regional laureate was Piotr Pawlikowski, in 2010 - Joanna Dziuba, in 2011 and 2012 - Paweł Rogaliński.[3] [4]

The excellent scientific staff of the higher education establishments in Łódź is complemented by Łódź's scientists from the Institute of the Polish Academy of Sciences (PAN) and scientific ministerial institutes working within the field of the occupational medicine, textile, paper and leather industries. The number of students in the higher education establishments in Lodz is still growing. Currently, they educate 113,000 students from Poland and other countries.

References

Notes

[1] Arkadiusz Belczyk, Tłumaczenie polskich nazw geograficznych na język angielski (http://serwistlumacza.com/content/view/27/32/) [Translation of Polish Geographical Names into English], 2002-2006.

[2] Stat.gov.pl (http://www.stat.gov.pl/gus/45_655_PLK_HTML.htm)

[3] http://www.studenckinobel.pl/index.php/historia-konkursu History of the contest "Studencki Nobel" (in Polish)

[4] "Młody dziennikarz znów pretenduje do Nobla! (in Polish)" (http://www.halolodz.pl/wiadomosci/wiadomoÅci/artykul,9369,mÅody-dziennikarz-znÃ³w-pretenduje-do-nobla.html). . Retrieved 2010-05-29.

External links

- Województwo Łódzkie (http://www.lodzkie.pl/) Official website
- www.lodzkie.travel – tourists attractions of łódź voivodeship (http://www.lodzkie.travel), a website produced by the Regional Tourist Organisation of the Łódź Voivodeship

Jogaila

Jogaila	
[[File:Jagiełło sarcophagus figure.jpg	Tomb effigy, Wawel Cathedral, Kraków]]
Tomb effigy, Wawel Cathedral, Kraków	
King of Poland	
Reign	4 March 1386-1 June 1434
Coronation	4 March 1386
Predecessor	Jadwiga
Successor	Władysław III
Spouse	Jadwiga of Poland Anne of Cilli Elisabeth of Pilica Sophia of Halshany
Issue	
Elizabeth Bonifacia Jadwiga of Lithuania Władysław III of Poland Casimir IV Jagiellon	
House	Jagiellon dynasty
Father	Algirdas
Mother	Uliana Alexandrovna of Tver
Born	about 1351/1362 Vilnius
Died	1 June 1434 Gródek Jagielloński (now Horodok, Ukraine)
Burial	Wawel Cathedral

Coat of Arms of Vladislav Jagaila as king of Poland

Poland and Lithuania 1386–1434

Jogaila, later **Władysław II Jagiełło**[1] (ca. 1351/1362 – 1 June 1434) was Grand Duke of Lithuania (1377–1434), jure uxoris King of Kingdom of Poland (1386–1399), and sole King of Poland (1399–1434). He ruled in Lithuania from 1377, at first with his uncle Kęstutis. In 1386 in Kraków he was baptized as Władysław, married the young queen regnant Jadwiga of Poland, and was crowned King of Poland as Władysław II Jagiełło.[2] In 1387 he converted Lithuania to Christianity. His own reign in Poland started in 1399, upon death of Queen Jadwiga, and lasted a further thirty-five years and laid the foundation for the centuries-long Polish–Lithuanian union. Władysław II was the founder of the Jagiellon dynasty that bears his name, while pagan Jogaila was an heir to the already established house of Gediminids in Grand Duchy of Lithuania; the royal dynasty ruled both states until 1572,[3] and became one of the most influential dynasties in the late medieval and early modern Central and Eastern Europe.[4] During his reign Polish-Lithuanian state was the greatest Kingdom of the Christian world[5].

Jogaila was the last pagan ruler of medieval Lithuania. After he became King of Poland, as a result of Union of Krewo, the newly formed Polish-Lithuanian union confronted the growing power of the Teutonic Knights. The allied victory at the Battle of Grunwald in 1410, followed by the Peace of Thorn (1411), secured the Polish and Lithuanian borders and marked the emergence of the Polish–Lithuanian alliance as a significant force in Europe. The reign of Władysław II Jagiełło extended Polish frontiers and is often considered the beginning of Poland's *Golden Age*.

Early life

Lithuania

Little is known of Jogaila's early life, and even his date of birth is not certain. Previously historians have given his date of birth as 1352, but some recent research suggests a later date—about 1362.[6] He was a descendant of the Gediminid dynasty and probably born in Vilnius. His parents were Algirdas, Grand Duke of Lithuania, and his second wife, Uliana, daughter of Alexander I, Grand Prince of Tver.

Grand Duchy of Lithuania to which Jogaila succeeded in 1377 was a political entity composed of two leading but very different nationalities and two political systems: ethnic Lithuania in the north-west and the vast Ruthenian territories of former Kievan Rus', comprising lands of modern Ukraine, Belarus, and parts of western Russia.[7] At first, Jogaila—like his father—based his rule in the southern and eastern territories of Lithuania, while his uncle,

Kęstutis, the Duke of Trakai, continued to rule the north-western region.[8] Jogaila's succession, however, soon placed this system of dual rule under strain.[4]

At the start of his reign, Jogaila was preoccupied with unrest in the Lithuanian Rus' lands. In 1377–78, Andrei of Polotsk, the eldest son of Algirdas, challenged Jogaila's authority and sought to become the Grand Duke. In 1380, Andrei and another brother, Dmitry, sided with Prince Dmitri of Moscow against Jogaila's alliance with emir Mamai, "de facto" khan of the Golden Horde.[9] Jogaila failed to support Mamai, lingering in the vicinity of the battlefield, which led to Mamai's army significant defeat at the hands of Prince Dmitri in the Battle of Kulikovo. The Muscovite quite Pyrrhic victory over the Golden Horde, in a long term, signified, however, the beginning of a slow climb to power by the Grand Duchy of Moscow, thus within a century becoming the most serious future rival and threat to integrity, well-being and survival of Lithuania. However, at that moment Muscovy was awfully weakened by tremendous losses suffered during the famous battle and thus in the same year Jogaila was free to begin a struggle for supremacy with Kęstutis.

In the north-west, Lithuania faced constant armed incursions from the Teutonic Knights—founded after 1226 to fight and convert the pagan Baltic tribes of Prussians, Yotvingians and Lithuanians. In 1380, Jogaila secretly concluded the secret Treaty of Dovydiškės, directed against Kęstutis.[4] When Kęstutis discovered the plan, the Lithuanian Civil War began. He seized Vilnius, overthrew Jogaila, and pronounced himself grand duke in his place.[10] In 1382, Jogaila raised an army from his father's vassals and confronted Kęstutis near Trakai. Kęstutis and his son Vytautas entered Jogaila's encampment for negotiations but were tricked and imprisoned in the Kreva Castle, where Kęstutis was found dead, probably murdered, a week later.[11] Vytautas escaped to the Teutonic fortress of Marienburg and was baptised there under the name Wigand.[10]

Jogaila formulated the Treaty of Dubysa, which rewarded the Knights for their aid in defeating Kęstutis and Vytautas by promising Christianisation and granting them Samogitia west of the Dubysa river. However, when Jogaila failed to ratify the treaty, the Knights invaded Lithuania in the summer of 1383. In 1384, Jogaila reconciled with Vytautas promising to return his patrimony in Trakai. Vytautas then turned against the Knights,attacking and looting several Prussian castles.[12]

Baptism and marriage

Jogaila's Russian mother Uliana of Tver urged him to marry Sofia, daughter of Prince Dmitri of Moscow, who required him first to convert to Orthodoxy.[13] That option, however, was unlikely to halt the crusades against Lithuania by the Teutonic Knights, who regarded Orthodox Christians as schismatics and little better than heathens.[4] [10] Jogaila chose therefore to accept a Polish proposal to become a Catholic and marry the eleven-year-old Queen Jadwiga of Poland.[14] The nobles of Malopolska made this offer to Jogaila for many reasons. For example, they wanted to neutralize the dangers from Lithuania itself and to secure the fertile territories of Galicia–Volhynia.[15] The Polish nobles saw the offer as an opportunity for increasing their privileges[16] and avoiding Austrian influence, brought by Jadwiga's previous fiancé William, Duke of Austria.[17]

On 14 August 1385 in Kreva Castle, Jogaila confirmed his prenuptial promises in the Union of Krewo (Union of Kreva). The promises included the adoption of Christianity, repatriation of lands "stolen" from Poland by its neighbours, and *terras suas Lithuaniae et Russiae Coronae Regni Poloniae perpetuo applicare*, a clause interpreted by historians to mean anything from a personal union between Lithuania and Poland to a complete incorporation of Lithuania into Poland.[18] The agreement at Kreva has been described both as far-sighted and as a desperate gamble.[19]

Jogaila was duly baptised at the Wawel Cathedral in Kraków on 15 February 1386 and from then on formally used the name Władysław or Latin versions of it.[20] [21] The marriage went ahead on 4 March 1386, two weeks after the baptism ceremonies, and Jogaila was crowned King Władysław by archbishop Bodzanta. He was also to be legally adopted by Jadwiga's mother, Elizabeth of Bosnia, so retaining the throne in the event of Jadwiga's death.[10] The royal baptism triggered the conversion of most of Jogaila's court and noblemen, as well as mass baptisms in

Lithuanian rivers,[22] a beginning of the final Christianization of Lithuania. Though the ethnic Lithuanian nobility were the main converts to Catholicism—both paganism and the Orthodox rite remained strong among the peasants—the king's conversion and its political implications created lasting repercussions for the history of both Lithuania and Poland.[22]

Ruler of Lithuania and Poland

Władysław II Jagiello and Queen Jadwiga reigned as co-monarchs; and though Jadwiga probably had little real power, she took an active part in Poland's political and cultural life. In 1387, she led two successful military expeditions to Red Ruthenia, recovered lands her father Louis I of Hungary had transferred from Poland to Hungary, and secured the homage of Petru I, Voivode of Moldavia.[23] In 1390, she also personally opened negotiations with the Teutonic Order. Most political responsibilities, however, fell to Jagiello, with Jadwiga attending to the cultural and charitable activities for which she is still revered.[23]

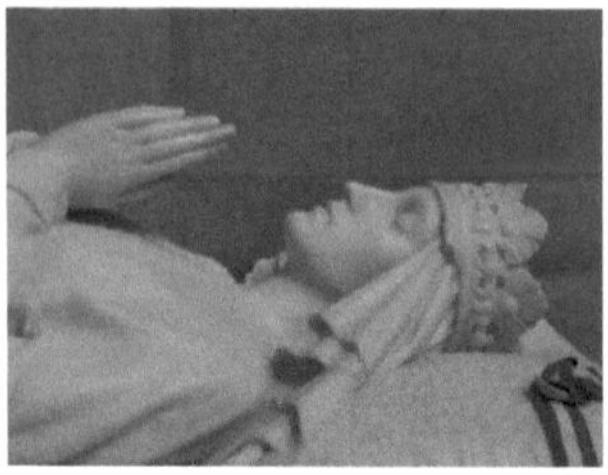

Jadwiga's sarcophagus, Wawel Cathedral

Soon after Jagiello's accession to the Polish throne, Jagiello granted Vilnius a city charter like that of Kraków, modeled on the Magdeburg Law; and Vytautas issued a privilege to a Jewish commune of Trakai on almost the same terms as privileges issued to the Jews of Poland in the reigns of Boleslaus the Pious and Casimir the Great.[24] Władysław's policy of unifying the two legal systems was partial and uneven at first but achieved a lasting influence.[23] By the time of the Union of Lublin in 1569, there was not much difference between the administrative and judicial systems in force in Lithuania and Poland.[25]

One effect of Jagiello's measures was to be the advancement of Catholics in Lithuania at the expense of Orthodox elements; in 1387 and 1413, for example, Lithuanian Catholic boyars were granted special judicial and political privileges denied to the Orthodox boyars.[26] As this process gained momentum, it was accompanied by the rise of both Rus' and Lithuanian identity in the fifteenth century.[27]

Challenges

Jagiello's baptism failed to end the crusade of the Teutonic Knights, who claimed his conversion was a sham, perhaps even a heresy, and renewed their incursions on the pretext that pagans remained in Lithuania.[10] [28] From then on, however, the Order found it harder to sustain the cause of a crusade and faced the growing threat to its existence posed by the Kingdom of Poland and a genuinely Christian Lithuania alliance.[29] [30] Władysław sponsored the creation of the diocese of Vilnius under bishop Andrzej Wasilko, the former confessor of Elisabeth of Hungary. The bishopric, which included Samogitia, then largely controlled by the Teutonic Order, was subordinated to the see of Gniezno and not to that of Teutonic Königsberg.[10] The decision may not have improved Władysław's relations with the Order, but it served to introduce closer ties between Lithuania and Poland, enabling the Polish church to freely assist its Lithuanian counterpart.[22]

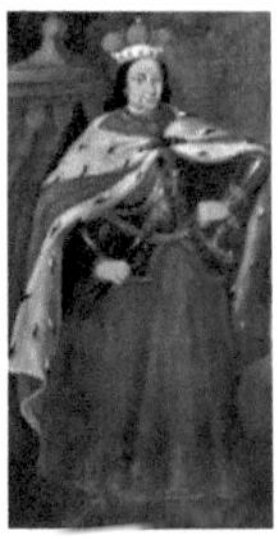

Vytautas the Great

In 1389, Władysław's rule in Lithuania faced a revived challenge from Vytautas, who resented the power given to Skirgaila in Lithuania at the expense of his own patrimony.[12] Vytautas started a civil war in Lithuania, aiming to become the Grand Duke. On 4 September 1390, the joint forces of Vytautas and the Teutonic Grand Master, Konrad von Wallenrode, laid siege to Vilnius, which was held by Władysław's regent Skirgaila with combined Polish, Lithuanian and Ruthenian troops.[4] Although the Knights lifted the siege of the castle after a month, they reduced much of the outer city to ruins. This bloody conflict was eventually brought to a temporary halt in 1392 with the Treaty of Ostrów, by which Władysław handed over the government of Lithuania to his cousin in exchange for peace: Vytautas was to rule Lithuania as the Grand Duke (*magnus dux*) until his death, under the overlordship of the Supreme Duke (*dux supremus*) in the person of the Polish monarch.[31] Skirgaila was moved from the Duchy of Trakai to become prince of Kiev.[32] Vytautas initially accepted his status but soon began to pursue Lithuania's independence from Poland.[23] [33]

The protracted period of war between the Lithuanians and the Teutonic Knights was ended on 12 October 1398 by the Treaty of Salynas, named after the islet in the Neman River where it was signed. Lithuania agreed to cede Samogitia and assist the Teutonic Order in a campaign to seize Pskov, while the Order agreed to assist Lithuania in a campaign to seize Novgorod.[23] Shortly afterwards, Vytautas was crowned as a king by local nobles; but the following year his forces and those of his ally, Khan Tokhtamysh of the White Horde, were crushed by the Timurids at the Battle of the Vorskla River, ending his imperial ambitions in the east and obliging him to submit to Władysław's protection once more.[4] [33]

King of Poland

On 22 June 1399, Jadwiga gave birth to a daughter, baptised Elizabeth Bonifacia; but within a month the mother and daughter died, leaving Władysław sole ruler of the Kingdom of Poland and without an heir nor much legitimacy to rule the kingdom. Jadwiga's death undermined Władysław's right to the throne; and as a result old conflicts between the nobility of Lesser Poland, generally sympathetic to Władysław, and the gentry of Greater Poland began to surface. In 1402, Władysław answered the rumblings against his rule by marrying Anna of Celje, a granddaughter of Casimir III of Poland, a political match which re-legitimised his reign.

The Union of Vilnius and Radom of 1401 confirmed Vytautas's status as grand duke under Władysław's overlordship, while assuring the title of grand duke to the heirs of Władysław rather than those of Vytautas: should Władysław die without heirs, the Lithuanian boyars were to elect a new monarch.[34] [35] Since no heir had yet been produced by either monarch, the act's implications were unforeseeable, but it forged bonds between the Polish and Lithuanian nobility and a permanent defensive alliance between the two states, strengthening Lithuania's hand for a new war against the Teutonic Order in which Poland officially took no part.[29] [33] While the document left the liberties of the Polish nobles untouched, it granted increased power to the boyars of Lithuania, whose grand dukes had till then been unencumbered by checks and balances of the sort attached to the Polish monarchy. The Union of Vilnius and Radom therefore earned Władysław a measure of support in Lithuania.[23]

In late 1401, the new war against the Order overstretched the resources of the Lithuanians, who found themselves fighting on two fronts after uprisings in the eastern provinces. Another of Władysław's brothers, the malcontent Švitrigaila, chose this moment to stir up revolts behind the lines and declare himself grand duke.[28] On 31 January 1402, he presented himself in Marienburg, where he won the backing of the Knights with concessions similar to those made by Jogaila and Vytautas during earlier leadership contests in the Grand Duchy.[34]

Against the Teutonic Order

Władysław II Jagiełło's royal seal

The war ended in the Treaty of Raciąż on 22 May 1404. Władysław acceded to the formal cession of Samogitia, and agreed to support the Order's designs on Pskov; in return, Konrad von Jungingen undertook to sell Poland the disputed Dobrzyń Land and the town of Złotoryja, once pawned to the Order by Władysław Opolski, and to support Vytautas in a revived attempt on Novgorod.[34] Both sides had practical reasons for signing the treaty at that point: the Order needed time to fortify its newly acquired lands, the Poles and Lithuanians to deal with territorial challenges in the east and in Silesia.

Also in 1404, Władysław held talks at Vratislav with Wenceslaus IV of Bohemia, who offered to return Silesia to Poland if Władysław supported him in his power struggle within the Holy Roman Empire.[36] Władysław turned the deal down with the agreement of both Polish and Silesian nobles, unwilling to burden himself with new military commitments in the west.[37]

Polish–Lithuanian–Teutonic war

Władysław II Jagiełło, by Marcello Bacciarelli

In December 1408, Władysław and Vytautas held strategic talks in Navahrudak Castle, where they decided to foment a Samogitian uprising against Teutonic rule to draw German forces away from Pomerelia. Władysław promised to repay Vytautas for his support by restoring Samogitia to Lithuania in any future peace treaty.[38] The uprising, which began in May 1409, at first provoked little reaction from the Knights, who had not yet consolidated their rule in Samogitia by building castles; but by June their diplomats were busy lobbying Władysław's court at Oborniki, warning his nobles against Polish involvement in a war between Lithuania and the Order.[39] Władysław, however, bypassed his nobles and informed new Grand Master Ulrich von Jungingen that if the Knights acted to suppress Samogitia, Poland would intervene. This stung the Order into issuing a declaration of war against Poland on 6 August, which Władysław received on 14 August in Nowy Korczyn.[39]

The castles guarding the northern border were in such bad condition that the Knights easily captured those at Złotoryja, Dobrzyń and Bobrowniki, the capital of Dobrzyń Land, while German burghers invited them into Bydgoszcz (German: Bromberg). Władysław arrived on the scene in late September, retook Bydgoszcz within a week, and came to terms with the Order on 8 October. During the winter, the two armies prepared for a major confrontation. Władysław installed a strategic supply depot at Płock in Masovia and had a pontoon bridge constructed and transported north down the Vistula.[40]

Meanwhile, both sides unleashed diplomatic offensives. The Knights dispatched letters to the monarchs of Europe, preaching their usual crusade against the heathens;[41] Władysław countered with his own letters to the monarchs, accusing the Order of planning to conquer the whole world.[42] Such appeals successfully recruited many foreign knights to each side. Wenceslas IV of Bohemia signed a defensive treaty with the Poles against the Teutonic Order; his brother, Sigismund of Luxembourg, allied himself with the Order and declared war against Poland on 12 July, though his Hungarian vassals refused his call to arms.[43]

Battle of Grunwald

Battle of Grunwald, 1410. Painting by Jan Matejko

When the war resumed in June 1410, Władysław advanced into the Teutonic heartland at the head of an army of about 20,000 mounted nobles, 15,000 armed commoners, and 2,000 professional cavalry mainly hired from Bohemia. After crossing the Vistula over the pontoon bridge at Czerwińsk, his troops met up with those of Vytautas, whose 11,000 light cavalry included Ruthenians and Tatars.[44] The Teutonic Order's army numbered about 18,000 cavalry, mostly Germans and 5,000 infantry. On 15 July, at the Battle of Grunwald (also known as Battle of Tannenberg) after one of the largest and most ferocious battles of the Middle Ages,[45] the allies won a victory so overwhelming that the Teutonic Order's army was virtually annihilated, with most of its key commanders killed in combat, including Grand Master Ulrich von Jungingen and Grand Marshal Friedrich von Wallenrode. Thousands of troops were reported to have been slaughtered on either side.[44]

The Teutonic Order's castle at Marienburg

The road to the Teutonic capital Marienburg now lay open, the city undefended; but for reasons the sources do not explain, Władysław hesitated to pursue his advantage.[46] On 17 July, his army began a laboured advance, arriving at Marienburg only on 25 July, by which time the new Grand Master, Heinrich von Plauen, had organised a defence of the fortress.[47] [48] The apparent half-heartedness of the ensuing siege, called off by Władysław on 19 September, has been ascribed variously to the impregnability of the fortifications,[] to high casualty figures among the Lithuanians, and to Władysław's unwillingness to risk further casualties; but a lack of sources precludes a definitive explanation.[49]

Final years

Dissent

Polish and Lithuanian conflict with Teutonic Prussia, 1377–1434.

The war ended in 1411 with the Peace of Thorn, in which neither Poland nor Lithuania drove home their negotiating advantage to the full, much to the discontent of the Polish nobles. Poland regained Dobrzyń Land, Lithuania regained Samogitia, and Masovia regained a small territory beyond the Wkra river. Most of the Teutonic Order's territory, however, including towns which had surrendered, remained intact. Władysław then proceeded to release many high-ranking Teutonic Knights and officials for apparently modest ransoms. The cumulative expense of the ransoms, however, proved a drain on the Order's resources.[50] This failure to exploit the victory to his nobles' satisfaction provoked growing opposition to Władysław's regime after 1411, further fuelled by the granting of Podolia, disputed between Poland and Lithuania, to Vytautas, and by the king's two-year absence in Lithuania.[51]

In an effort to outflank his critics, Władysław promoted the leader of the opposing faction, bishop Mikołaj Trąba, to the archbishopric of Gniezno in autumn 1411 and replaced him in Kraków with Wojciech Jastrzębiec, a supporter of Vytautas.[51] He also sought to create more allies in Lithuania. In the Union of Horodło, signed on 2 October 1413, he decreed that the status of the Grand Duchy of Lithuania was "tied to our Kingdom of Poland permanently and irreversibly" and granted the Catholic nobles of Lithuania privileges equal to those of the Polish szlachta. The act included a clause prohibiting the Polish nobles from electing a monarch without the consent of the Lithuanian nobles, and the Lithuanian nobles from electing a grand duke without the consent of the Polish monarch.[7] [52]

Last conflicts

In 1414, a sporadic new war broke out, known as the "Hunger War" from the Knights' scorched-earth tactics of burning fields and mills; but both the Knights and the Lithuanians were too exhausted from the previous war to risk a major battle, and the fighting petered out in the autumn.[51] Hostilities did not flare up again until 1419, during the Council of Constance, when they were called off at the papal legate's insistence.[51]

The Council of Constance proved a turning point in the Teutonic crusades, as it did for several European conflicts. Vytautas sent a delegation in 1415, including the metropolitan of Kiev; and Samogitian witnesses arrived at Constance at the end of that year to point out their preference for being "baptised with water and not with blood".[53] The Polish envoys, among them Mikołaj Trąba, Zawisza Czarny, and Paweł Włodkowic, lobbied for an end to the forced conversion of heathens and to the Order's aggression against Lithuania and Poland.[54] As a result of the Polish–Lithuanian diplomacy, the council, though scandalised by Włodkowic's questioning of the monastic state's legitimacy, denied the Order's request for a further crusade and instead entrusted the conversion of the Samogitians to Poland–Lithuania.[55]

The diplomatic context at Constance included the revolt of the Bohemian Hussites, who looked upon Poland as an ally in their wars against Sigismund, the emperor elect and new king of Bohemia. In 1421, the Bohemian Diet declared Sigismund deposed and formally offered the crown to Władysław on condition he accept the religious principles of the Four Articles of Prague, which he was not prepared to do. After Władysław's refusal, Vytautas was postulated (elected in absentia) as Bohemian king, but he assured the pope that he opposed the heretics. Between 1422 and 1428, Władysław's nephew, Sigismund Korybut, attempted a regency in war-torn Bohemia, with little success.[56] Vytautas accepted Sigismund's offer of a royal crown in 1429—apparently with Władysław's blessing—but Polish forces intercepted the crown in transit and the coronation was cancelled.[7] [57]

In 1422, Władysław fought another war, known as the Gollub War, against the Teutonic Order, defeating them in under two months before the Order's imperial reinforcements had time to arrive. The resulting Treaty of Melno ended the Knights' claims to Samogitia once and for all and defined a permanent border between Prussia and Lithuania. Lithuania was given the province of Samogitia, with the port of Palanga, but the city of Klaipėda was left to the Order.[7] This border remained largely unchanged for roughly 500 years, until 1920. The terms of this treaty have, however, been seen as turning a Polish victory into defeat, as a result of Władysław's renunciation of Polish claims to Pomerania, Pomerelia, and Chełmno Land, for which he received only the town of Nieszawa in return.[58] The Treaty of Melno closed a chapter in the Knights' wars with Lithuania but did little to settle their long-term issues with Poland. Further sporadic warfare broke out between Poland and the Knights between 1431 and 1435.

Cracks in the cooperation between Poland and Lithuania after the death of Vytautas in 1430 had offered the Knights a revived opportunity for interference in Poland. Władysław supported his brother Švitrigaila as grand duke of Lithuania,[59] but when Švitrigaila, with the support of the Teutonic Order and dissatisfied Rus' nobles,[9] rebelled against Polish overlordship in Lithuania, the Poles, under the leadership of Bishop Zbigniew Oleśnicki of Kraków, occupied Podolia, which Władysław had awarded to Lithuania in 1411, and Volhynia.[35] In 1432, a pro-Polish party in Lithuania elected Vytautas's brother Žygimantas as grand duke,[59] leading to an armed struggle over the Lithuanian succession which stuttered on for years after Władysław's death.[9] [7]

Succession

Władysław's second wife, Anna of Celje, had died in 1416, leaving a daughter, Jadwiga. In 1417, Władysław married Elisabeth of Pilica, who died in 1420 without bearing him a child, and two years later, Sophia of Halshany, who bore him two surviving sons. The death in 1431 of Princess Jadwiga, the last heir of Piast blood, released Władysław to make his sons by Sophia of Halshany his heirs, though he had to sweeten the Polish nobles with concessions to ensure their agreement, since the monarchy was elective. Władysław finally died in 1434, leaving Poland to his elder son, Władysław III, and Lithuania to his younger, Casimir, both still minors at the time.[60] [61] The Lithuanian inheritance, however, could not be taken for granted. His death in 1434 ended the personal union between the two realms, and it was not clear what would take its place.[62]

Family tree (incomplete)

See also

- History of Poland (1385–1569)
- Jogaila (Władysław II Jagiello): names and titles
- List of Belarusian rulers
- List of Lithuanian rulers
- King Jagiello Monument
- List of Poles

Notes

[1] He is known under a number of names: Lithuanian: *Jogaila Algirdaitis*; Polish: *Władysław II Jagiełło*; Belarusian: Jahajła (Ягайла). See also: Jogaila (Władysław II Jagiello): names and titles.

[2] Bojtár 2000, p. 182

[3] Anna Jagiellon last member of royal Jagiellon family died in 1596.

[4] Bojtár 2000, pp. 180–186

[5] Anna Boczkowska, Sarkofag Władysława II Jagiełły i Donatello, Gdańsk 2011, p. 27.

[6] Tęgowski 1999, pp. 124–125

[7] Stone 2001, p. 4

[8] Some historians have called this system a diarchy (Sruogienė-Sruoga 1987; Deveike 1950). However, Rowell suggests that the nature of this dual rule "...reflects political expediency; it certainly does not meet the formal definition of diarchy as 'rule by two independent authorities'...those two leaders were not equal: the grand duke in Vilnius was supreme" (Rowell 1994, p. 68).

[9] Plokhy 2006, p. 46

[10] Rowell 2000, pp. 709–712

[11] Bojtár 2000, p. 181

[12] Mickūnaitė 1998, p. 157

[13] The historian John Meyendorff suggests Jogaila may have already been an Orthodox Christian: "In 1377, Olgerd of Lithuania died, leaving the Grand Principality to his son Jagiello, an Orthodox Christian..." (Meyendorff 1989, p. 205). Dmitri, however, made it a condition of the marriage that Jogaila "should be baptized in the Orthodox faith and that he should proclaim his Christianity to all men" (Dvornik 1992, p. 221).

[14] Jadwiga had actually been crowned king of Poland (*rex poloni*), because the Polish political system made no provision for a queen-regent Stone 2001, p. 8).

[15] Lukowski & Zawadzki 2001, p. 42

[16] Dvornik 1992, p. 129

[17] Lukowski & Zawadzki 2001, p. 37

[18] Lukowski & Zawadzki 2001, p. 41; Stone 2001, p. 8

[19] It "reflects the exceptional far-sightedness of the political elites ruling both countries" (Kłoczowski 2000, p. 55). It was "a desperate gamble by Jogaila to avert a seemingly inevitable subjugation" (Lukowski & Zawadzki 2001, p. 38)

[20] Sruogienė-Sruoga 1987

[21] A Slavic name that roughly translates as *glorious rule*, Władysław is often Latinised into either Wladislaus or Ladislaus. The choice evoked both Władysław I of Poland, the Elbow-high, who was Queen Jadwiga's great-grandfather and unified the kingdom in 1320, and Saint Ladislaus I of Hungary, a king who sided with the pope against the emperor Henry IV and Christianised Transylvania (Rowell 2000,

pp. 709–712).

[22] Kłoczowski 2000, pp. 54–57

[23] Jasienica 1988, pp. 80–146

[24] Jasienica 1988, pp. 74–80

[25] Dvornik 1992, p. 344

[26] Magocsi 1996, p. 134

[27] Plokhy 2006, p. 98

[28] Housley 1992, p. 354

[29] Sedlar 1994, p. 388

[30] Turnbull 2004, p. 22

[31] Rowell 2000, p. 732

[32] Stone 2001, p. 10

[33] Dvornik 1992, pp. 222–225

[34] Jasienica 1988, pp. 103–105

[35] Stone 2001, p. 11

[36] *New Cambridge Medieval History*, 348.

[37] "Śląsk w polityce Piastów (*Silesia within the policies of the Piasts*)" (http://www.piastowie.kei.pl/piast2/slask.htm). *Polska Piastów*. 2005. . Retrieved 2006-08-09.

[38] Karwasińska & Zakrzewski 1935, p. 21

[39] Jasienica 1988, pp. 106–107

[40] Turnbull 2003, pp. 32–33

[41] Delbrück 1990, p. 526

[42] Jasienica 1988, p. 108

[43] Jasienica 1988, p. 110

[44] Stone 2001, p. 16

[45] Bojtár 2000, p. 182; Turnbull 2003, p. 7

[46] Turnbull 2003, p. 7

[47] Stone 2001, p. 17

[48] Turnbull 2003, p. 73

[49] Jasienica,, pp. 113–120

[50] *New Cambridge Medieval History*, 364.

[51] Jasienica 1988, pp. 121–124

[52] Dvornik 1992, pp. 342–343; *New Cambridge Medieval History*, 775–6.

[53] Housley 1992, p. 361; Rowell 2000, p. 733

[54] Kłoczowski 2000, p. 73

[55] Housley 1992, pp. 351–361

[56] Bideleux 1998, pp. 233–235; Turnbull & McBride 2004, pp. 11–12

[57] *New Cambridge Medieval History*, 353.

[58] Jasienica 1988, p. 130

[59] Sruogienė-Sruoga 1987

[60] Sedlar 1994, p. 282

[61] Rowell 2001, p. 711

[62] Stone 2001, p. 22

[63] Ryszard Jurzak, *Władysław II Jagiełło*, in *Genealogy*, (2006). (http://genealog.home.pl/) Retrieved 2006-09-27.

Footnotes

References

- Bideleux, Robert (1998), *A History of Eastern Europe: Crisis and Change*, Routledge, ISBN 0-415-16111-8
- Bojtár, Endre (2000), *Foreword to the Past: A Cultural History of the Baltic People*, translated by Walter J. Renfroe, Budapest: Central European University Press, ISBN 963-9116-42-4
- Delbrück, Hans (1990), *The Barbarian Invasions: History of the Art of War* (http://books.google.com/?id=hdh4Elj-3WEC&pg=PA526), University of Nebraska Press, ISBN 0-8032-6447-X
- Deveike, Jone (1950), "The Lithuanian Diarchies", *The Slavonic and East European Review* **28** (71): 392–405

- Długosz, Jan (1997), *The Annals of Jan Dlugosz*, translated and abridged by Maurice Michael, IM Publications, ISBN 1-901019-00-4
- Drabina, Jan (1994), "Die Religionspolitik von König Wladyslaw Jagiello im polnisch-litauischen Reich in den Jahren 1385–1434", *Zeitschrift für Ostforschung* **43** (2): 161–173, ISSN 0044-3239 (German)
- Dvornik, Francis (1992), *The Slavs in European History and Civilization*, Rutgers University Press, ISBN 0-8135-0799-5
- Housley, Norman (1992), *The Later Crusades 1274–1580: From Lyons to Alcazar*, Oxford University Press, ISBN 0-19-822136-3
- Jakštas, J.; Ivinskis, Z.; Sužiedėlis, S.; Šapoka, A.; Šležas, P. (1935), *Jogaila*, Kaunas: redagavo A. Šapoka (Lithuanian)
- Jasienica, Paweł (1988), *Polska Jagiellonów*, Warsaw: Państwowy Instytut Wydawniczy, ISBN 83-06-01796-X (Polish)
- Jurzak, Ryszard (2006), "Władysław II Jagiełło" (http://genealog.home.pl/), *Dynastic Genealogy*, retrieved 27 September 2006
- Karwasińska, Jadwiga & Zakrzewski, Ignacy (1935), *Lites ac res gestae inter Polonos Ordinemque Cruciferorum* (http://books.google.com/?id=TfQKAAAAIAAJ), Warsaw: Bibliotheca Cornicensis (Polish) (Latin)
- Kłoczowski, Jerzy (2000), *A History of Polish Christianity* (http://books.google.com/?id=ecdye8hk_tgC&pg=PA55), Cambridge University Press, ISBN 0-521-36429-9
- Kosman, Marceli (1976), *Drogi zaniku pogaństwa u Bałtów*, Wrocław: Zakład Narodowy im. Ossolińskich (Polish)
- Kosman, Marceli (1987), *Jagiełło królem Polski: z dziejów unii Polski i Litwy*, Warsaw: Wydawnictwa Szkolne i Pedagogiczne (WSiP), ISBN 83-02-03292-1 (Polish)
- Lukowski, Jerzy & Zawadzki, Hubert (2001), *A Concise History of Poland*, Cambridge University Press, ISBN 0-521-55917-0
- Magocsi, Paul Robert (1996), *A History of Ukraine*, University of Toronto Press, ISBN 0-8020-7820-6
- Meyendorff, John (1989), *Byzantium and the Rise of Russia* (Reprint ed.), Crestwood, NY: Vladimir's Seminary Press, ISBN 0-88141-079-9
- Mickūnaitė, Giedrė (1999), "From Pamphlet to Origin Theory" (http://books.google.com/books?vid=ISBN9042008342&id=Nl0_nXQD5i4C&pg=PA157&sig=DoKb27Z8XzT8ZZP3qQnwuL3-pqw), in Kooper, Erik Simon, *The Medieval Chronicle: Proceedings of the 2nd International Conference on the Medieval Chronicle*, Amsterdam: Rodopi, ISBN 90-420-0834-2
- O'Connor, Kevin (2003), *A History of the Baltic States*, Greenwood Press, ISBN 0-313-32355-0
- Plokhy, Serhii (2006), *The Origins of the Slavic Nations: Premodern Identities in Russia, Ukraine, and Belarus*, Cambridge University Press, ISBN 0-521-86403-8
- Rogalski, Leon (1846), *Dzieje Krzyżaków (History of the Teutonic Knights*, Warsaw: Samuel Orgelbrand (Polish)
- Rowell, S. C. (2000), "Baltic Europe", in Jones, Michael, *The New Cambridge Medieval History VI*, Cambridge University Press, ISBN 0-521-36290-3
- Rowell, S. C. (1994), *Lithuania Ascending: A Pagan Empire Within East-central Europe, 1295–1345*, Cambridge University Press, ISBN 0-521-45011-X
- Sedlar, Jean W. (1994), *East Central Europe in the Middle Ages, 1000–1500* (http://books.google.com/?id=ANdbpi1WAIQC&pg=PA388), Seattle: University of Washington Press, ISBN 0-295-97290-4
- *Śląsk w polityce Piastów* (http://www.piastowie.kei.pl/piast2/slask.htm), Polska Piastów, 2005, retrieved 9 August 2006 (Polish)
- Sruogienė-Sruoga, Vanda (1987), "Jogaila (1350–1434)" (http://www.lituanus.org/1987/87_4_04.htm), *Lituanus: Lithuanian Quarterly Journal of Arts and Sciences* **33** (4), retrieved 25 January 2007
- Stone, Daniel Z. (2001), *The Polish-Lithuanian State, 1386–1795*, Seattle: University of Washington Press, ISBN 0-295-98093-1

- Tęgowski, Jan (1999), *Pierwsze pokolenia Giedyminowiczow*, Poznań-Wrocław: Wydawn. Historyczne, ISBN 83-913563-1-0 (**Polish**)
- Turnbull, Stephen (2004), *Crusader Castles of the Teutonic Knights (2): Baltic Stone Castles 1184–1560* (http://books.google.com/?id=F3wzkSHR4j4C&pg=PA22), Osprey Publishing, ISBN 1-84176-712-3
- Turnbull, Stephen (2003), *Tannenberg 1410: Disaster for the Teutonic Knights*, Osprey, ISBN 1-84176-561-9
- Turnbull, Stephen & McBride, Angus (2004), *The Hussite Wars: 1419–36*, Osprey Publishing, ISBN 1-84176-665-8

External links

- "Wladyslaw II" at the Directory of Royal Genealogical Data (http://www3.dcs.hull.ac.uk/cgi-bin/gedlkup/n=royal?royal05545)

Poland

Republic of Poland

Rzeczpospolita Polska

Anthem:

Mazurek Dąbrowskiego
(*Poland Is Not Yet Lost*)

Location of Poland(dark green)

– in Europe(green & dark grey)
– in the European Union(green) — [Legend]

Capital (and largest city)	Warsaw 52°13′N 21°02′E
Official language(s)	Polish
Regional language(s)	Kashubian
Ethnic groups (2011)	91.56% Polish 0.94% Silesian 0.04% Kashubian 0.07% German 0.07% Ukrainian 0.08% Belarusian 7.24% Other[1]
Demonym	Pole/Polish
Government	Parliamentary republic
- President	Bronisław Komorowski
- Prime Minister	Donald Tusk
Legislature	National Assembly

-	Upper house	Senate
-	Lower house	Sejm
Formation		
-	Christianisation[c]	April 14, 966
-	Kingdom of Poland	April 18, 1025
-	Polish-Lithuanian commonwealth	July 1, 1569
-	Partition of Poland	October 24, 1795
-	Duchy of Warsaw	July 22, 1807
-	Congress Poland	June 9, 1815
-	Reconstitution of Poland	November 11, 1918
-	Invasion of Poland, World War II	September 1, 1939
-	Communist Poland	April 8, 1945
-	Republic of Poland	September 13, 1989 –
Area		
-	Total	312685 km^2 [d](69th) 120,696.41 sq mi
-	Water (%)	3.07
Population		
-	2010 estimate	38,186,860[2] (34th)
-	2011 census	38,501,000
-	Density	120/km^2 (83rd) 319.9/sq mi
GDP (PPP)		2011 estimate
-	Total	$771.658 billion[3]
-	Per capita	$20,334[3]
GDP (nominal)		2011 estimate
-	Total	$513.821 billion[3]
-	Per capita	$13,540[3]
Gini (2002)		34.5 (low)
HDI (2011)		▲0.813[4] (very high) (39th)
Currency		*Złoty* (PLN)
Time zone		CET (UTC+1)
-	Summer (DST)	CEST (UTC+2)
Drives on the		right
Calling code		48
ISO 3166 code		PL
Internet TLD		.pl
1.	**a** See, however, Unofficial mottos of Poland.	

2.	**b** Although not official languages, Belarusian, Kashubian, Lithuanian and German are used in 20 communal offices.
3.	**c** The adoption of Christianity in Poland is seen by many Poles, regardless of their religious affiliation or lack thereof, as one of the most significant national historical events; the new religion was used to unify the tribes in the region.
4.	**d** The area of Poland according to the administrative division, as given by the Central Statistical Office, is 312679 km^2 (**unknown operator: u'strong'** sq mi) of which 311888 km^2 (**unknown operator: u'strong'** sq mi) is land area and 791 km^2 (**unknown operator: u'strong'** sq mi) is internal water surface area.[5]

Poland i/ˈpoʊlənd/ (Polish: *Polska*), officially the **Republic of Poland** (Polish: *Rzeczpospolita Polska*; Kashubian: *Pòlskô Repùblika*), is a country in Central Europe, bordered by Germany to the west; the Czech Republic and Slovakia to the south; Ukraine, Belarus and Lithuania to the east; and the Baltic Sea and Kaliningrad Oblast, a Russian exclave, to the north. The total area of Poland is 312679 square kilometres (**unknown operator: u'strong'** sq mi),[5] making it the 69th largest country in the world and the 9th largest in Europe. Poland has a population of over 38.5 million people,[5] which makes it the 34th most populous country in the world[6] and the sixth most populous member of the European Union, being its most populous post-communist member. Poland is a unitary state made up of 16 voivodeships. Poland is a member of the European Union, NATO, the United Nations, the World Trade Organization, the Organisation for Economic Co-operation and Development (OECD), European Economic Area, International Energy Agency, Council of Europe, Organization for Security and Co-operation in Europe, International Atomic Energy Agency, European Space Agency, G6, Council of the Baltic Sea States, Visegrád Group, Weimar Triangle and Schengen Agreement.

The establishment of a Polish state is often identified with the adoption of Christianity by its ruler Mieszko I in 966,[7] over the territory similar to that of present-day Poland. The Kingdom of Poland was formed in 1025, and in 1569 it cemented a long association with the Grand Duchy of Lithuania by signing the Union of Lublin, forming the Polish–Lithuanian Commonwealth. The Commonwealth ceased to exist in 1795 as the Polish lands were partitioned among the Kingdom of Prussia, the Russian Empire, and Old Austria. Poland regained its independence as the Second Polish Republic in 1918. Two decades later, in September 1939, World War II started with the Nazi Germany and Soviet Union invasion of Poland (Molotov–Ribbentrop Pact). Over six million Polish citizens died in the war. The People's Republic was declared in 1952 although Poland was a client state of the Soviet Union from 1944. During the Revolutions of 1989, the communist state was overthrown and democratic rule was re-established in the form of the current Poland, constitutionally known as the "Third Polish Republic".

Despite the vast destruction the country experienced in World War II, Poland managed to preserve much of its cultural wealth. There are currently 14 heritage sites inscribed on the UNESCO World Heritage list in Poland.[8] Since the end of the communist period, Poland has achieved a "very high" ranking in terms of human development.[9]

Etymology

The source of the name Poland[10] and the ethnonyms for the Poles[11] include endonyms (the way Polish people refer to themselves and their country) and exonyms (the way other peoples refer to the Poles and their country). Endonyms and most exonyms for Poles and Poland derive from the name of the West Slavic tribe of the Polans (*Polanie*).

The origin of the name *Polanie* itself is uncertain. It may derive from such Polish words as *pole* (field).[12] The early tribal inhabitants denominated it from the nature of the country. Lowlands and low hills predominate throughout the vast region from the Baltic shores to the foothills of the Carpathian Mountains. *Inter Alpes Huniae et Oceanum est Polonia, sic dicta in eorum idiomate quasi Campania* is the description by Gervase of Tilbury in his *Otia imperialia* (Recreation for the emperor, 1211). In some languages the exonyms for Poland derive from another tribal name, Lechites (*Lechici*).

History

Prehistory

Map of Poland (Latin: *Polonie*) in 960-992 under Mieszko, her first crowned King

Historians have postulated that throughout Late Antiquity, many distinct ethnic groups populated the regions of what is now known as Poland. The ethnicity and linguistic affiliation of these groups have been hotly debated; the time and route of the original settlement of Slavic peoples in these regions have been the particular subjects of much controversy.

The most famous archeological find from the prehistory and protohistory of Poland is the Biskupin fortified settlement (now reconstructed as a museum), dating from the Lusatian culture of the early Iron Age, around 700 BC. Before adopting Christianity in 960 AD, the people of Poland believed in Svetovid, the Slavic god of war, fertility, and abundance. Many other Slavic nations had the same belief.

Piast dynasty

Baptism of Poland in 966 AD.

Poland began to form into a recognizable unitary and territorial entity around the middle of the 10th century under the Piast dynasty. Poland's first historically documented ruler, Mieszko I, was baptized in 966, adopting Catholicism as the nation's new official religion, to which the bulk of the population converted in the course of the next few centuries. In 1000 Boleslaw the Brave, continuing the policy of his father held a Congress of Gniezno and created a new dioceses (Gniezno, Kraków, Kołobrzeg, Wrocław).

In 1109 Bolesław III Wrymouth defeated in Battle of Hundsfeld King of Germany Henry V. In 1138, Poland fragmented into several smaller duchies when Bolesław III Wrymouth divided the nation amongst his sons. In 1226 Konrad I of Masovia, one of the regional Piast dukes, invited the Teutonic Knights to help him fight the Baltic Prussian pagans; a decision which would ultimately lead to centuries of warfare with the Knights. In the middle of 13th-century Poland was almost united by Silesian branch of Piast dynasty (Henry I the Bearded and Henry II the Pious), when the country was devastated by the Mongols and the Battle of Legnica where Duke Henry II the Pious died. In 1320, after a number of earlier unsuccessful attempts by regional rulers at uniting the Polish dukedoms, Władysław I consolidated his power, took the throne and became the first King of a reunified Poland. His son, Casimir III, is remembered as one of the greatest Polish kings, who was widely recognized as a protector of trade. He extended his kingdom to 250% of its initial size. Casimir is also known for extending royal protection to Jews and providing the original impetus for the establishment of Poland's first university. The Golden Liberty of the nobles began to develop under Casimir's rule, when in return for their military support, the king made serious concessions to the aristocrats, finally establishing their status as superior to that of the townsmen, and aiding their rise to power. When Casimir died in 1370 he left no legitimate male heir and, considering his other male descendants either too young or unsuitable, was laid to rest as the last of the nation's Piast rulers.

Poland was also a centre of migration of peoples. The Germans settled in the towns, the Jewish community began to settle and flourish in Poland during this era (see History of the Jews in Poland); the same applies in smaller number to Armenians. The Black Death which affected most parts of Europe from 1347 to 1351 affected Poland to lesser extent.[13]

Jagiellon dynasty

Nicolaus Copernicus 'talking with God', who reveals the truth about the solar system to him. Painting by Jan Matejko.

The rule of the Jagiellon dynasty spanned the late Middle Ages and early Modern Era of Polish history. Beginning with the Lithuanian Grand Duke Jogaila (Władysław II Jagiełło), the Jagiellon dynasty (1386–1572) formed the Polish–Lithuanian union. The partnership brought vast Lithuania-controlled Rus' areas into Poland's sphere of influence and proved beneficial for the Poles and Lithuanians, who coexisted and cooperated in one of the largest political entities in Europe for the next four centuries. In the Baltic Sea region Poland's struggle with the Teutonic Knights continued and included the Battle of Grunwald (1410), where a Polish-Lithuanian army inflicted a decisive defeat on the Teutonic Knights, both countries' main adversary, allowing Poland's and Lithuania's territorial expansion into the far north region of Livonia.[14] In 1466, after the Thirteen Years' War, King Casimir IV Jagiellon gave royal consent to the milestone Peace of Thorn, which created the future Duchy of Prussia, a Polish vassal. The Jagiellons at one point also established dynastic control over the kingdoms of Bohemia (1471 onwards) and Hungary.[15] [16] In the south Poland confronted the Ottoman Empire and the Crimean Tatars (by whom they were attacked on 75 separate occasions between 1474 and 1569),[17] and in the east helped Lithuania fight the Grand Duchy of Moscow. Some historians estimate that Crimean Tatar slave-raiding cost Poland one million of its population from 1494 to 1694.[18]

Wawel, the seat of Polish kings. Kraków was the nation's capital from 1038 until the move to Warsaw in 1596

Poland was developing as a feudal state, with a predominantly agricultural economy and an increasingly powerful landed nobility. The *Nihil novi* act adopted by the Polish Sejm (parliament) in 1505, transferred most of the legislative power from the monarch to the Sejm, an event which marked the beginning of the period known as "Golden Liberty", when the state was ruled by the "free and equal" Polish nobility. Protestant Reformation movements made deep inroads into Polish Christianity, which resulted in the establishment of policies promoting religious tolerance, unique in Europe at that time. It is believed that this tolerance allowed the country to avoid the religious turmoil that spread over Europe during the late Middle Ages. The European Renaissance evoked in late Jagiellon Poland (kings Sigismund I the Old and Sigismund II Augustus) a sense of urgency in the need to promote a cultural awakening, and resultantly during this period Polish culture and the nation's economy flourished. In 1543 the Pole, Nicolaus Copernicus, an astronomer from Toruń, published his epochal works, *De revolutionibus orbium coelestium* (*On the Revolutions of the Celestial Spheres*), and thus became the first proponent of a predictive mathematical model confirming heliocentric theory which ultimately became the accepted basic model for the practice of modern astronomy. Another major figure associated with the era is classicist poet Jan Kochanowski.[19]

Polish–Lithuanian Commonwealth

The 1569 Union of Lublin established the Polish–Lithuanian Commonwealth, a more closely unified federal state with an elective monarchy, but which was governed largely by the nobility, through a system of local assemblies with a central parliament. The establishment of the Commonwealth coincided with a period of great stability and prosperity in Poland, with the union soon thereafter becoming a great European power and a major cultural entity, occupying approximately one million square kilometres of central Europe, as well as an agent for the of the 'Western culture' through Polonization in modern-day Ukraine, Belarus and Western Russia. Poland-Lithuania suffered from a number of dynastic crises during the reigns of the Vasa kings Sigismund III and Władysław IV and found itself engaged in a major conflicts with Russia, Sweden and the Ottoman Empire, as well as a series of minor Cossack uprisings.[20]

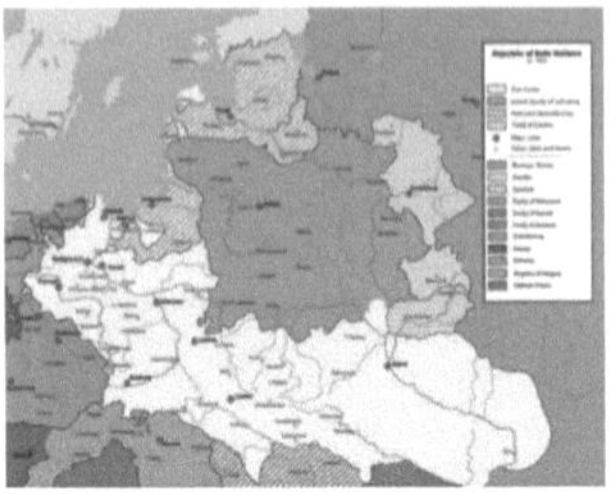

The Polish–Lithuanian Commonwealth at its greatest extent, after the Truce of Deulino (Dywilino) of 1619

A stylised depiction of King John III Sobieski, allied commander-in-chief and victor at the Battle of Vienna

From the middle of the 17th century, the nobles' democracy, suffering from internal disorder, gradually declined, thus leaving the once powerful Commonwealth extremely vulnerable to foreign intervention. From 1648, the Cossack Khmelnytsky Uprising engulfed the south and east eventually leaving Ukraine divided, with the eastern part, lost by the Commonwealth, becoming a dependency of the Tsardom of Russia. This was soon followed by the 'Deluge', a Swedish invasion, which raged through the Polish heartlands and caused unprecedented damage to Poland's population, culture and infrastructure. Famines and epidemics followed hostilities, and the population dropped from roughly 11 to 7 million.[21] However, under John III Sobieski the Commonwealth's military prowess was re-established, and in 1683 Polish forces played a major part in relieving Vienna of a major Turkish siege which was being conducted by Kara Mustafa in hope of eventually marching his troops further into Europe to spread Islam. Unfortunately, Sobieski's reign was to mark the end of the nation's golden-era, and soon, finding itself subjected to almost constant warfare and suffering enormous population losses as well as massive damage to its economy, the Commonwealth fell into decline. The government became ineffective as a result of large scale internal conflicts (e.g. Lubomirski's Rokosz against John II Casimir and rebellious confederations) and corrupted legislative processes. The nobility fell under the control of a handful of *magnats*, and this, compounded with two relatively weak kings of the Saxon Wettin dynasty, Augustus II and Augustus III, as well as the rise of Russia and Prussia after the Great Northern War only served to worsen the Commonwealth's plight. Despite this The Commonwealth-Saxony personal union gave rise to the emergence of the Commonwealth's first reform movement, and laid the foundations for the Polish Enlightenment.[22]

During the later part of the 18th century, the Commonwealth made attempts to implement fundamental internal reforms; with the second half of the century bringing a much improved economy, significant population growth and far-reaching progress in the areas of education, intellectual life, art, and especially toward the end of the period, evolution of the social and political system. The most populous capital city of Warsaw replaced Gdańsk (Danzig) as the leading centre of commerce, and the role of the more prosperous townsfolk soon increased. The royal election of 1764 resulted in the elevation of Stanisław August Poniatowski, a refined and worldly aristocrat connected to a major magnate faction, to the monarchy. However, a one-time lover of Empress Catherine II of Russia, the new king spent much of his reign torn between his desire to implement reforms necessary to save his nation, and his perceived necessity to remain in a relationship with his Russian sponsor. This ultimately led to the formation of the 1768 Bar Confederation; a *szlachta* rebellion directed against Russia and the Polish king which fought to preserve Poland's independence and the *szlachta*'s traditional privileges. Attempts at reform provoked the union's neighbours, and in 1772 the First Partition of the Commonwealth by Russia, Austria and Prussia took place; an act which the "Partition Sejm", under considerable duress, eventually "ratified" *fait accompli.*[23] Disregarding this loss, in 1773 the king established the Commission of National Education, the first government education authority in Europe.

Stanisław August Poniatowski, the last King of Poland and Grand Duke of Lithuania acceded to the throne in 1764, reigning until his abdication on 25 November 1795

The long-lasting Great Sejm convened by Stanisław August in 1788 successfully adopted the May 3 Constitution, the first set of modern supreme national laws in Europe. However, this document, accused by detractors of harbouring revolutionary sympathies, soon generated strong opposition from the Commonwealth's nobles and conservatives as well as from Catherine II, who, determined to prevent the rebirth of a strong Commonwealth set about planning the final dismemberment of the Polish-Lithuanian state. Russia was greatly aided in achieving its goal when the Targowica Confederation, an organisation of Polish nobles, appealed to the Empress for help, and in May 1792 Russian forces crossed the Commonwealth's frontier, thus beginning the Polish-Russian War. The defensive war fought by the Poles and Lithuanians ended prematurely when the King, convinced of the futility of resistance, capitulated and joined the Targowica Confederation. The Confederation then took over the government; Russia and Prussia, fearing the mere existence of a Polish state, arranged for, and subsequently in 1793, executed the Second Partition of the Commonwealth, which left the country deprived of so much territory that it was practically incapable of independent existence. Eventually, in 1795, following the failed Kościuszko Uprising, the Commonwealth was partitioned one last time by all three of its more powerful neighbours, and with this, effectively ceased to exist.[24]

The Age of Partitions

Tadeusz Kościuszko takes the oath to the King on the Rynek in Kraków, 1794

Poles rebelled several times against the partitioners, particularly near the end of the 18th century and the beginning of the 19th century. One of the most famous and successful attempts at securing renewed Polish independence took place in 1794, during the Kościuszko Uprising, at the Racławice where Tadeusz Kosciuszko, a popular and distinguished general who had served under Washington in America, led peasants and some Polish regulars into battle against numerically superior Russian forces. In 1807, Napoleon I of France recreated a Polish state, the Duchy of Warsaw, but after the Napoleonic Wars, Poland was again divided by the victorious Allies at the Congress of Vienna of 1815. The eastern part was ruled by the Russian tsar as a Congress Kingdom which possessed a very liberal constitution. However, the tsars soon reduced Polish freedoms, and Russia annexed the country in virtually all but name. Thus in the latter half of the 19th century, only Austrian-ruled Galicia, and particularly the Free City of Kraków, created good environment for free Polish cultural life to flourish.

Throughout the period of the partitions, political and cultural repression of the Polish nation led to the organisation of a number of uprisings against the authorities of the occupying Russian, Prussian and Austrian governments. Notable amongst these are the November Uprising of 1830 and January Uprising of 1863, both of which were attempts to free Poland from the rule of tsarist Russia. The November uprising began on 29 November 1830 in Warsaw when, led by Lieutenant Piotr Wysocki, young non-commissioned officers at the Imperial Russian Army's military academy in that city revolted. They were soon joined by large segments of Polish society, and together forced Warsaw's Russian garrison to withdraw north of the city.

Polish insurgents and Russian cuirassiers clash on a bridge in Warsaw's Łazienki Park during the November Uprising, Painting by Wojciech Kossak, 1898

Over the course of the next seven months, Polish forces successfully defeated the Russian armies of Field Marshal Hans Karl von Diebitsch and a number of other Russian commanders; however, finding themselves in a position unsupported by any other foreign powers, save distant France and the newborn United States, and with Prussia and Austria refusing to allow the import of military supplies through their territories, the Poles accepted that the uprising was doomed to failure. Upon the surrender of Warsaw to General Ivan Paskievich, many Polish troops, feeling they could not go on, withdrew into Germany and there laid down their arms. Poles would have to wait another 32 years for another opportunity to free their homeland.

When in January 1863 a new Polish uprising against Russian rule began, it did so as a spontaneous protest by young Poles against conscription into the Imperial Russian Army. However, the insurrectionists, despite being joined by high-ranking Polish-Lithuanian officers and numerous politicians were still severely outnumbered and lacking in foreign support. They were forced to resort to guerrilla warfare tactics and ultimately failed to win any major military victories. Afterwards no major uprising was witnessed in the Russian controlled Congress Poland and Poles resorted instead to fostering economic and cultural self-improvement.

Despite the political unrest experienced during the partitions, Poland did benefit from large scale industrialisation and modernisation programs, instituted by the occupying powers, which helped it develop into a more economically coherent and viable entity. This was particularly true in the Greater Poland, Pomerania and Warmia annexed by Prussia (later becoming a part of the German Empire); an area which eventually, thanks largely to the Greater Poland

Uprising, was reconstituted as a part of the Second Polish Republic and became one of its most productive regions.

Reconstitution of Poland

Chief of State Marshal Józef Piłsudski

During World War I, all the Allies agreed on the reconstitution of Poland that United States President Woodrow Wilson proclaimed in Point 13 of his Fourteen Points. A total of 2 million Polish troops fought with the armies of the three occupying powers, and 450,000 died.[25] Shortly after the armistice with Germany in November 1918, Poland regained its independence as the Second Polish Republic (*II Rzeczpospolita Polska*). It reaffirmed its independence after a series of military conflicts, the most notable being the Polish–Soviet War (1919–1921) when Poland inflicted a crushing defeat on the Red Army at the Battle of Warsaw, an event which is considered to have ultimately halted the advance of Communism into Europe and forced Vladimir Lenin to rethink his objective of achieving global socialism. Nowadays the event is often referred to as the 'Miracle at the Vistula'.[26]

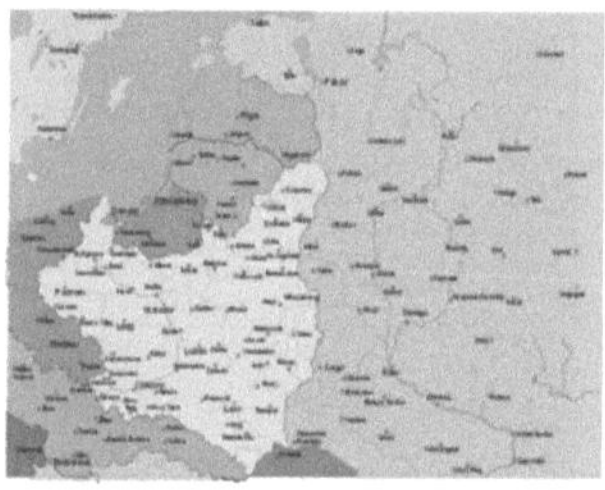

Poland between 1922 and 1938

During this period, Poland successfully managed to fuse the territories of the three former partitioning powers into a cohesive nation state. Railways were restructured to direct traffic towards Warsaw instead of the former imperial capitals, a new network of national roads was gradually built up and a major seaport was opened on the Baltic Coast, so as to allow Polish exports and imports to bypass the politically charged Free City of Danzig.

The inter-war period heralded in a new era of Polish politics. Whilst Polish political activists had faced heavy censorship in the decades up until the First World War, the country now found itself trying to establish a new political tradition. For this reason, many exiled Polish activists, such as Jan Paderewski (who would later become Prime Minister) returned home to help; a great number of them then went on to take key positions in the newly formed political and governmental structures. Tragedy struck in 1922 when Gabriel Narutowicz, inaugural holder of the Presidency, was assassinated at the Zachęta Gallery in Warsaw by painter and right-wing nationalist Eligiusz Niewiadomski.[27]

The 1926 May Coup of Józef Piłsudski turned rule of the Second Polish Republic over to the Sanacja movement. By the 1930s Poland had become increasingly authoritarian; a number of 'undesirable' political parties, such as the Polish Communists, had been banned and following Piłsudski's death, the regime, unable to appoint a new leader, began to show its inherent internal weaknesses and unwillingness to cooperate in any way with other political parties.

World War II

The Sanacja movement controlled Poland until the start of World War II in 1939, when Nazi Germany's and Slovakia Invasion of Poland (1939) on 1 September and the Soviet invasion of Poland on 17 September, which followed the breaking of the Soviet–Polish Non-Aggression Pact, occurred. Warsaw capitulated on 28 September 1939. As agreed in the Molotov–Ribbentrop Pact, Poland was split into two zones, one occupied by Nazi Germany while the Kresy, or Borderlands, fell under the control of the Soviet Union. In 1939-1941, the Soviets had moved hundreds of thousands of Poles across the Soviet Union, and the Soviet secret police, NKVD, had executed thousands of Polish prisoners of war (inter alia Katyn massacre).[28]

Polish forces stationed abroad constituted the fourth largest allied force of the war

Poland made the fourth-largest troop contribution to the Allied war effort, after the Soviets, the British and the Americans.[a] Polish troops fought under the command of both the Polish Government in Exile in the theatre of war west of Germany and under Soviet leadership in the theatre of war east of Germany. The Polish expeditionary corps, which was controlled by the exiled pre-war government based in London, played an important role in the Italian and North African Campaigns.[29] [30] They are particularly well remembered for their conduct at the Battle of Monte Cassino, a conflict which culminated in the raising of a Polish flag over the ruins of the mountain-top abbey by the 12th Podolian Uhlans. The Polish forces in the theatre of war east of Germany were commanded by Lieutenant General Władysław Anders who had received his command from Prime Minister of the exiled government Władysław Sikorski. On the east of Germany, the Soviet-backed Polish 1st Army distinguished itself in the battles for Berlin and Warsaw, although its actions in support of the latter have often been criticised.

At the end of World War II, the gray territories were transferred from Poland to the Soviet Union, and the pink territories from Germany to Poland

Polish servicemen were also active in the theatres of naval and air warfare; during the Battle of Britain Polish squadrons such as the No. 303 "Kościuszko" fighter squadron[31] achieved great success, and by the end of the war the exiled Polish Air Forces could claim 769 confirmed kills. Meanwhile, the Polish Navy was active in the protection of convoys in the North Sea and Atlantic Ocean.[32]

In addition to the organised units of the 1st Army and the Forces in the Nazi-occupied Europe, the domestic underground resistance movement, the Armia Krajowa, or *Home Army*, fought to free Poland from German occupation and establish an independent Polish state. The wartime resistance movement in Poland was one of the three largest resistance movements of the entire war[b] and encompassed an unusually broad range of clandestine activities, which essentially functioned as an underground state complete with degree-awarding universities and a court system.[33] The resistance was, however, largely loyal to the exiled government and generally resented the idea of a communist Poland; for this reason, on 1 August 1944 they initiated Operation Tempest and thus began the Warsaw Uprising.[34] [35] The objective of the uprising was to drive the German occupiers from the city and help with the larger fight against Germany and the Axis powers, however secondary motives for the uprising sought to see Warsaw liberated before the Soviets could reach the capital, so as to underscore Polish sovereignty by empowering the Polish Underground State before the Soviet-backed Polish Committee of National Liberation could assume control. However, a lack of available allied military aid and Stalin's reluctance to allow the 1st Army to help their fellow countrymen take the city, ultimately led to the uprising's failure and subsequent planned destruction of the city.

Grave of Polish fighter killed during the Warsaw Uprising. The battle lasted 63 days, and resulted in the deaths of 200,000 people.

During the war, German forces, under direct order from Adolf Hitler, set up six major extermination camps, all of which were established on Polish territory; these included both the notorious Treblinka and Auschwitz camps. This allowed the Germans to transport the Jews living in the Third Reich outside of "German" territory, as well as to import Jews and other targeted groups from across occupied Europe to be "liquidated," or killed, in the concentration camps set up in the General Government. Among such groups were Polish intelligentsia, communists, Roma peoples and Soviet Prisoners of War. However, since millions of Jews lived in pre-war Poland, Jewish victims make up the largest percentage of all victims of the Nazis' extermination program. It is estimated that, of pre-war Poland's Jewry, approximately 90% (or about 3 million members) were killed. Throughout the occupation, many members of the Armia Krajowa, supported by the Polish government in exile, and millions of ordinary Poles — at great risk to themselves and their families — engaged in rescuing Jews from the Nazi Germans. Grouped by nationality, Poles represent the largest number of people who rescued Jews during the Holocaust.[36] [37] To date, 6,135 Poles have been awarded the title of *Righteous among the Nations* by the State of Israel–more than any other nation.[36] Some estimates put the number of Poles involved in rescue efforts at up to 3 million, and credit Poles with saving up to approximately 450,000 Jews from certain death.[37]

At the war's conclusion, Poland's territory was shifted westwards, pushing the Kresy in accordance with the Curzon Line. Meanwhile, the western border moved to the Oder-Neisse line. As a result, Poland's territory was reduced by 20%, or 77500 square kilometres (**unknown operator: u'strong'** sq mi). The shift forced the migration of millions of people, most of whom were Poles, Germans, Ukrainians, and Jews.[38] Of all the countries involved in the war, Poland lost the highest percentage of its citizens: over 6 million perished — nearly one-fifth of Poland's population — half of them Polish Jews. Over 90% of deaths were non-military in nature. Only in the 1970s did Poland again approach its prewar population levels.

Postwar communist Poland

At the insistence of Joseph Stalin, the Yalta Conference sanctioned the formation of a new Polish provisional and pro-Communist coalition government in Moscow, which ignored the Polish government-in-exile based in London; a move which angered many Poles who considered it a betrayal by the Allies. In 1944, Stalin had made guarantees to Churchill and Roosevelt that he would maintain Poland's sovereignty and allow democratic elections to take place; however, upon achieving victory in 1945, the occupying Soviet authorities organised an election which constituted nothing more than a sham and was ultimately used to claim the 'legitimacy' of Soviet hegemony over Polish affairs. The Soviet Union instituted a new communist government in Poland, analogous to much of the rest of the Eastern Bloc. As elsewhere in Communist Europe the Soviet occupation of Poland met with armed resistance from the outset which continued into the fifties.

At High Noon, June 4, 1989 – political poster featuring Gary Cooper to encourage votes for the Solidarity (Polish trade union)Solidarity party in the Contract Sejm1989 elections. "The 4th of June, 1989 marked a decisive victory for democracy in Poland and, ultimately, across (Central and) Eastern Europe.— Angela Merkel (English) "Merkel honours Polish freedom struggle and Tiananmen victims". www.thelocal.de. AFP. . Retrieved 2009-09-28.

Despite widespread objections, the new Polish government accepted the Soviet annexation of the pre-war eastern regions of Poland[40] (in particular the cities of Wilno and Lwów) and agreed to the permanent garrisoning of Red Army units on Poland's territory. Military alignment within the Warsaw Pact throughout the Cold War came about as a direct result of this change in Poland's political culture and in the European scene came to characterise the fully-fledged integration of Poland into the brotherhood of communist nations.

The People's Republic of Poland (*Polska Rzeczpospolita Ludowa*) was officially proclaimed in 1952. In 1956 after the death of Bolesław Bierut, the régime of Władysław Gomułka became temporarily more liberal, freeing many people from prison and expanding some personal freedoms. A similar situation repeated itself in the 1970s under Edward Gierek, but most of the time persecution of anti-communist opposition groups persisted. Despite this, Poland was at the time considered to be one of the least repressive states of the Soviet Bloc.[41]

Gomułka with Leonid Brezhnev in East Germany

Labour turmoil in 1980 led to the formation of the independent trade union "Solidarity" ("*Solidarność*"), which over time became a political force. Despite persecution and imposition of martial law in 1981, it eroded the dominance of the Communist Party and by 1989 had triumphed in Poland's first partially free and democratic parliamentary elections since the end of the Second World War. Lech Wałęsa, a Solidarity candidate, eventually won the presidency in 1990. The Solidarity movement heralded the collapse of communist regimes and parties across Europe.

Present-day Poland

Poland joined NATO in 1999 and since 2004 has been a member of the European Union.

A shock therapy programme, initiated by Leszek Balcerowicz in the early 1990s enabled the country to transform its socialist-style planned economy into a market economy. As with all other post-communist countries, Poland suffered temporary slumps in social and economic standards, but it became the first post-communist country to reach its pre-1989 GDP levels, which it achieved by 1995 largely thanks to its booming economy.[42] [43]

Most visibly, there were numerous improvements in human rights, such as the freedom of speech, civil liberties (1st class) and political rights (1st class), according to Freedom House. In 1991, Poland became a member of the Visegrád Group and joined the North Atlantic Treaty Organization (NATO) alliance in 1999 along with the Czech Republic and Hungary. Poles then voted to join the European Union in a referendum in June 2003, with Poland becoming a full member on 1 May 2004. Subsequently Poland joined the Schengen Area in 2007, as a result of which, the country's borders with other member states of the European Union have been dismantled, allowing for full freedom of movement within most of the EU.[44] In contrast to this, the section of Poland's eastern border now comprising the external EU border with Belarus, Russia and Ukraine, has become increasingly well protected, and has led in part to the coining of the phrase 'Fortress Europe', in reference to the seeming 'impossibility' of gaining entry to the EU for citizens of the former Soviet Union.

On April 10, 2010, the President of the Republic of Poland, Lech Kaczyński, along with 89 other high-ranking Polish officials died in a plane crash near Smolensk, Russia. The president's party were on their way to attend an annual service of commemoration for the victims of the Katyń massacre when the tragedy took place.

Geography

Poland's territory extends across several geographical regions, between latitudes 49° and 55° N, and longitudes 14° and 25° E. In the north-west is the Baltic seacoast, which extends from the Bay of Pomerania to the Gulf of Gdańsk. This coast is marked by several spits, coastal lakes (former bays that have been cut off from the sea), and dunes. The largely straight coastline is indented by the Szczecin Lagoon, the Bay of Puck, and the Vistula Lagoon. The centre and parts of the north lie within the North European Plain.

Poland's topography

Rising gently above these lowlands is a geographical region comprising the four hilly districts of moraines and moraine-dammed lakes formed during and after the Pleistocene ice age. These lake districts are the Pomeranian Lake District, the Greater Polish Lake District, the Kashubian Lake District, and the Masurian Lake District. The Masurian Lake District is the largest of the four and covers much of north-eastern Poland. The lake districts form part of the Baltic Ridge, a series of moraine belts along the southern shore of the Baltic Sea.

South of the Northern European Lowlands lie the regions of Silesia and Masovia, which are marked by broad ice-age river valleys. Farther south lies the Polish mountain region, including the Sudetes, the Cracow-Częstochowa Upland, the Świętokrzyskie Mountains, and the Carpathian Mountains, including the Beskids. The highest part of the Carpathians is the Tatra Mountains, along Poland's southern border.

Geology

Granite outcrop Silesian Rocks at Karkonosze (Giant Mountains), south-western Poland

The geological structure of Poland has been shaped by the continental collision of Europe and Africa over the past 60 million years, on the one hand (and the other), by the Quaternary glaciations of northern Europe. Both processes shaped the Sudetes and the Carpathian Mountains. The moraine landscape of northern Poland contains soils made up mostly of sand or loam, while the ice age river valleys of the south often contain loess. The Cracow-Częstochowa Upland, the Pieniny, and the Western Tatras consist of limestone, while the High Tatras, the Beskids, and the Karkonosze are made up mainly of granite and basalts. The Polish Jura Chain is one of the oldest mountain ranges on earth.

Poland has 70 mountains over 2,000 metres (6,600 ft) in elevation, all in the Tatras. The Polish Tatras, which consist of the High Tatras and the Western Tatras, is the highest mountain group of Poland and of the entire Carpathian range. In the High Tatras lies Poland's highest point, the north-western peak of Rysy, 2499 metres (8199 ft) in elevation. At its foot lies the mountain lakes of Czarny Staw pod Rysami (Black Lake below Mount Rysy), and Morskie Oko (the Marine Eye).

Giewont in the Tatra Mountains; the mountainous south is a popular destination for hikers

The second highest mountain group in Poland is the Beskids, whose highest peak is Babia Góra, at 1725 metres (5659 ft). The next highest mountain groups is the Karkonosze in the Sudetes, whose highest point is Śnieżka, at 1602 metres (5256 ft); Śnieżnik Mountains whose highest point is Śnieżnik, at 1425 metres (4675 ft).

Tourists also frequent the Bieszczady Mountains in the far southeast of Poland, whose highest point in Poland is Tarnica, with an elevation of 1346 metres (4416 ft), Gorce Mountains in Gorce National Park, whose highest point is Turbacz, with elevations 1310 metres (4298 ft), and the Pieniny in Pieniny National Park, whose highest point is Wysokie Skałki (Wysoka), with elevations 1050 metres (3445 ft). The lowest point in Poland — at 2 metres (6.6 ft) below sea level — is at Raczki Elbląskie, near Elbląg in the Vistula Delta.

Dunes in Słowiński National Park

The only desert located in Poland stretches over the Zagłębie Dąbrowskie (the Coal Fields of Dąbrowa) region. It is called the Błędów Desert, located in the Silesian Voivodeship in southern Poland. It has a total area of 32 square kilometres (12 sq mi). It is one of only five natural deserts in Europe. But also, it is the warmest desert that appears at this latitude. Błędów Desert was created thousands of years ago by a melting glacier. The specific geological structure has been of big importance. The average thickness of the sand layer is about 40 metres (131 ft), with a maximum of 70 metres (230 ft), which made the fast and deep drainage very easy.

The Baltic Sea activity in Słowiński National Park created sand dunes which in the course of time separated the bay from the sea. As waves and wind carry sand inland the dunes slowly move, at a speed of 3 to 10 metres (9.8 to 32.8 ft) meters per year. Some dunes are quite high – up to 30 metres (98 ft). The highest peak of the park — Rowokol (115 metres / 377 feet above sea level) — is also an excellent observation point.

Waters

Vistula River in Modlin

The longest rivers are the Vistula (Polish: *Wisła*), 1047 kilometres (**unknown operator: u'strong'** mi) long; the Oder (Polish: *Odra*) which forms part of Poland's western border, 854 kilometres (**unknown operator: u'strong'** mi) long; its tributary, the Warta, 808 kilometres (**unknown operator: u'strong'** mi) long; and the Bug, a tributary of the Vistula, 772 kilometres (**unknown operator: u'strong'** mi) long. The Vistula and the Oder flow into the Baltic Sea, as do numerous smaller rivers in Pomerania.

The Łyna and the Angrapa flow by way of the Pregolya to the Baltic, and the Czarna Hańcza flows into the Baltic through the Neman. While the great majority of Poland's rivers drain into the Baltic Sea, Poland's Beskids are the source of some of the upper tributaries of the Orava, which flows via the Váh and the Danube to the Black Sea. The eastern Beskids are also the source of some streams that drain through the Dniester to the Black Sea.

Poland's rivers have been used since early times for navigation. The Vikings, for example, traveled up the Vistula and the Oder in their longships. In the Middle Ages and in early modern times, when the Polish–Lithuanian Commonwealth was the breadbasket of Europe; the shipment of grain and other agricultural products down the Vistula toward Gdańsk and onward to other parts of Europe took on great importance.

With almost ten thousand closed bodies of water covering more than 1 hectare (2.47 acres) each, Poland has one of the highest numbers of lakes in the world. In Europe, only Finland has a greater density of lakes. The largest lakes, covering more than 100 square kilometres (39 sq mi), are Lake Śniardwy and Lake Mamry in Masuria, and Lake Łebsko and Lake Drawsko in Pomerania.

Kurtkowiec, oligotrophic lake in southeastern Poland

In addition to the lake districts in the north (in Masuria, Pomerania, Kashubia, Lubuskie, and Greater Poland), there is also a large number of mountain lakes in the Tatras, of which the Morskie Oko is the largest in area. The lake with the greatest depth—of more than 100 metres (328 ft)—is Lake Hańcza in the Wigry Lake District, east of Masuria in Podlaskie Voivodeship.

Holidaymakers relax at the Lake Solina near Lesko in southeastern Poland

Among the first lakes whose shores were settled are those in the Greater Polish Lake District. The stilt house settlement of Biskupin, occupied by more than one thousand residents, was founded before the 7th century BC by people of the Lusatian culture.

Lakes have always played an important role in Polish history and continue to be of great importance to today's modern Polish society. The ancestors of today's Poles, the Polanie, built their first fortresses on islands in these lakes. The legendary Prince Popiel is supposed to have ruled from Kruszwica on Lake Gopło. The first historically documented ruler of Poland, Duke Mieszko I, had his palace on an island in the Warta River in Poznań. Nowadays the Polish lakes provide an invaluable location for the pursuit of water sports such as yachting and wind-surfing.

The Polish Baltic coast is approximately 528 kilometres (328 mi) long and extends from Świnoujście on the islands of Usedom and Wolin in the west to Krynica Morska on the Vistula Spit in the east. For the most part, Poland has a smooth coastline, which has been shaped by the continual movement of sand by currents and winds. This continual erosion and deposition has formed cliffs, dunes, and spits, many of which have migrated landwards to close off former lagoons, such as Łebsko Lake in Słowiński National Park.

Bay of Puck *(Zatoka Pucka)* in Poland

Prior to the end of the Second World War and subsequent change in national borders, Poland had only a very small coastline; this was situated at the end of the 'Polish Corridor', the only internationally recognised Polish territory which afforded the country access to the sea. However after World War II, the redrawing of Poland's borders and resulting 'shift' of the country's borders left it with a greatly expanded coastline, thus allowing for far greater access to the sea than was ever previously possible. The significance of this event, and importance of it to Poland's future as a major industrialised nation, was allured to by the 1945 Wedding to the Sea.

The largest spits are Hel Peninsula and the Vistula Spit. The largest Polish Baltic island is Wolin. The largest port cities are Gdynia, Gdańsk, Szczecin, and Świnoujście. The main coastal resorts are Sopot, Międzyzdroje, Kołobrzeg, Łeba, Władysławowo, and the Hel Peninsula.

Land use

Forests of the Ojców National Park

Poland is the fourth most forested country in Europe. Forests cover about 30,5% of Poland's land area based on international standards. Its overall percentage is still increasing. Forests of Poland is managed by the national program of reforestation (KPZL), aiming at an increase of forest-cover to 33% in 2050. The richness of Polish forest (per SoEF 2011 statistics) is more than twice as high as European average (with Germany and France at the top), containing 2.304 billion cubic metres of trees.[45] The largest forest complex in Poland is Lower Silesian Wilderness.

More than 1% of Poland's territory, 3145 square kilometres (**unknown operator: u'strong'** sq mi), is protected within 23 Polish national parks. Three more national parks are projected for Masuria, the Kraków-Częstochowa Upland, and the eastern Beskids. In addition, wetlands along lakes and rivers in central Poland are legally protected, as are coastal areas in the north. There are over 120 areas designated as landscape parks, along with numerous nature reserves and other protected areas (e.g. Natura 2000).

Present day Poland is a country with great agricultural prospects; there are over two million private farms in the country, and Poland is the leading producer in Europe of potatoes and rye and is one of the world's largest producers of sugar beets and triticale. This has led Poland to be described on occasion as the future 'bread basket of the European Union'. However, despite employing around 16% of the workforce, agricultural output in Poland remains low and the industry is characterised as largely inefficient due to the large number of small, independent farms. This situation is likely to soon change for the better with the government debating agricultural reform and currently pursuing the option of auctioning off large tracts of state-owned agricultural land.

Biodiversity

Phytogeographically, Poland belongs to the Central European province of the Circumboreal Region within the Boreal Kingdom. According to the World Wide Fund for Nature, the territory of Poland can be subdivided into three ecoregions: the Baltic mixed forests, Central European mixed forests and Carpathian montane conifer forests.

Family of White Stork. Poland hosts the largest White Stork population.[46]

A wisent in the Białowieża Forest

Many animals that have since died out in other parts of Europe still survive in Poland, such as the wisent in the ancient woodland of the Białowieża Forest and in Podlaskie. Other such species include the brown bear in Białowieża, in the Tatras, and in the Beskids, the gray wolf and the Eurasian Lynx in various forests, the moose in northern Poland, and the beaver in Masuria, Pomerania, and Podlaskie.

In the forests, one also encounters game animals, such as red deer, roe deer and wild boars. In eastern Poland there are a number of ancient woodlands, like Białowieża forest, that have never been cleared by people. There are also large forested areas in the mountains, Masuria, Pomerania, Lubusz Land and Lower Silesia.

Poland is the most important breeding ground for European migratory birds. Out of all of the migratory birds who come to Europe for the summer, one quarter breed in Poland, particularly in the lake districts and the wetlands along the Biebrza, the Narew, and the Warta, which are part of nature reserves or national parks.

Climate

The climate is mostly temperate throughout the country. The climate is oceanic in the north and west and becomes gradually warmer and continental towards the south and east. Summers are generally warm, with average temperatures between 18 °C (**unknown operator: u'strong'** °F) and 30 °C (86.0 °F) depending on a region. Winters are rather cold, with average temperatures around 3 °C (37.4 °F) in the northwest and −6 °C (**unknown operator: u'strong'** °F) in the northeast. Precipitation falls throughout the year, although, especially in the east; winter is drier than summer.

The average daytime summer temperature at sea level along the Baltic coast is 22 °C (71.6 °F)[47]

The warmest region in Poland is Lower Silesian located in south-western Poland where temperatures in the summer average between 24 °C (75.2 °F) and 32 °C (**unknown operator: u'strong'** °F) but can go as high as 34 °C (93.2 °F) to 39 °C (102.2 °F) on some days in the warmest month of July and August. The warmest cities in Poland are Tarnów, which is situated in Lesser Poland and Wrocław, which is located in Lower Silesian. The average temperatures in Wrocław are 20 °C (**unknown operator: u'strong'** °F) in the summer and 0 °C (32.0 °F) in the winter, but Tarnów has the longest summer in all of Poland, which lasts for 115 days, from mid-May to mid-September. The coldest region of Poland is in the northeast in the Podlaskie Voivodeship near the border of Belarus. Usually the coldest city is Suwałki. The climate is affected by cold fronts which come from Scandinavia and Siberia. The average temperature in the winter in Podlaskie ranges from −6 °C (**unknown operator: u'strong'** °F) to −4 °C (**unknown operator: u'strong'** °F).

Politics

President of Poland Bronisław Komorowski

Poland is a democracy, with a president as a head of state, whose current constitution dates from 1997. The government structure centers on the Council of Ministers, led by a prime minister. The president appoints the cabinet according to the proposals of the prime minister, typically from the majority coalition in the Sejm. The president is elected by popular vote every five years. The president is Bronisław Komorowski. Komorowski replaced President Lech Kaczyński following the latter's death in an April 10, 2010 air crash. The prime minister, Donald Tusk, was appointed in 2007 after his Civic Platform party made significant gains in that year's parliamentary elections. In 2011, Tusk became the first Polish prime minister in history to be democratically re-elected for a consecutive term.

Polish voters elect a bicameral parliament consisting of a 460-member lower house (Sejm) and a 100-member Senate (Senat). The Sejm is elected under proportional representation according to the d'Hondt method, a method similar to that used in many parliamentary political systems. The Senat, on the other hand, is elected under the First-past-the-post voting method, with one senator being returned from each of the 100 constituencies.

The Sejm's session chamber in Warsaw

With the exception of ethnic minority parties, only candidates of political parties receiving at least 5% of the total national vote can enter the Sejm. When sitting in joint session, members of the Sejm and Senat form the National Assembly (the *Zgromadzenie Narodowe*). The National Assembly is formed on three occasions: when a new President takes the oath of office; when an indictment against the President of the Republic is brought to the State Tribunal (*Trybunał Stanu*); and when a president's permanent incapacity to exercise his duties because of the state of his health is declared. To date only the first instance has occurred.

The judicial branch plays an important role in decision-making. Its major institutions include the Supreme Court of the Republic of Poland (*Sąd Najwyższy*); the Supreme Administrative Court of the Republic of Poland (*Naczelny Sąd Administracyjny*); the Constitutional Tribunal of the Republic of Poland (*Trybunał Konstytucyjny*); and the State Tribunal of the Republic of Poland (*Trybunał Stanu*). On the approval of the Senat, the Sejm also appoints the ombudsman or the Commissioner for Civil Rights Protection (*Rzecznik Praw Obywatelskich*) for a five-year term. The ombudsman has the duty of guarding the observance and implementation of the rights and liberties of Polish citizens and residents, of the law and of principles of community life and social justice.

In 2011, Poles elected Anna Grodzka as the first ever transsexual MP in European history,[48] and the second transgender MP in European history, after the Italian Vladimir Luxuria.[49]

Law

The Supreme Court building in Warsaw

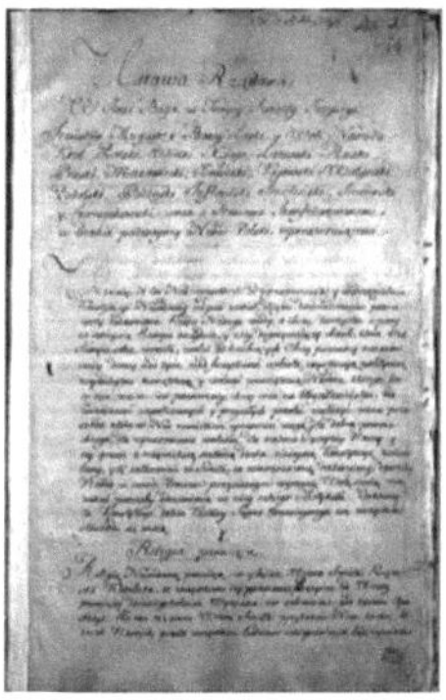

The Constitution of May 3, 1791 has been called the first of its kind in Europe.[50]

The Constitution of Poland is the supreme law in contemporary Poland, and the Polish legal system is based on the principle of civil rights, governed by the code of Civil Law. Historically, the most famous Polish legal act is the Constitution of May 3, 1791. Historian Norman Davies describes it as the first of its kind in Europe.[50] The Constitution was instituted as a Government Act (Polish: *Ustawa rządowa*) and then adopted on May 3, 1791 by the Sejm of the Polish–Lithuanian Commonwealth. Primarily, it was designed to redress long-standing political defects of the federative Polish–Lithuanian Commonwealth and its Golden Liberty. Previously only the Henrican articles signed by each of Poland's elected kings could perform the function of a set of basic laws. The new Constitution introduced political equality between townspeople and the nobility (*szlachta*), and placed the peasants under the protection of the government, thus mitigating the worst abuses of serfdom. The Constitution abolished pernicious parliamentary institutions such as the *liberum veto*, which at one time had placed the sejm at the mercy of any deputy who might choose, or be bribed by an interest or foreign power, to have rescinded all the legislation that had been passed by that sejm. The May 3rd Constitution sought to supplant the existing anarchy fostered by some of the country's reactionary magnates, with a more egalitarian and democratic constitutional monarchy.

Unfortunately, the adoption of such a liberal constitution was treated as a grave threat by Poland's more autocratic neighbours. In response Prussia, Austria and Russia formed an anti-Polish alliance and over the next decade collaborated with one another to partition their weaker neighbour and ultimately destroy the Polish state. In the words of two of its co-authors, Ignacy Potocki and Hugo Kołłątaj, the constitution represented "the last will and testament of the expiring Fatherland." Despite this, its text influenced many later democratic movements across the globe.

Poland's current constitution was adopted by the National Assembly of Poland on 2 April 1997, approved by a national referendum on 25 May 1997, and came into effect on 17 October 1997. It guarantees a multi-party state, the freedoms of religion, speech and assembly, and specifically casts off many Communist ideals to create a 'free market economic system'. It requires public officials to pursue ecologically sound public policy and acknowledges the inviolability of the home, the right to form trade unions, and to strike, whilst at the same time prohibiting the practices of forced medical experimentation, torture and corporal punishment.

Foreign relations

Foreign Secretary Sikorski meets US Secretary of State Hillary Clinton

In recent years, Poland has extended its responsibilities and position in European and international affairs, supporting and establishing friendly relations with other European nations and a large number of 'developing' countries.

In 1994, Poland became an associate member of the European Union (EU) and its defensive arm, the Western European Union (WEU), having submitted preliminary documentation for full membership in 1996, it formally joined the European Union in May 2004, along with the other members of the Visegrád group. In 1996, Poland achieved full OECD membership, and at the 1997 Madrid Summit was invited to join the North Atlantic Treaty Organisation (NATO) in the first wave of policy enlargement finally becoming a full member of NATO in March 1999.

As changes since the fall of Communism in 1989 have redrawn the map of central Europe, Poland has tried to forge strong and mutually beneficial relationships with its seven new neighbours, this has notably included signing 'friendship treaties' to replace links severed by the collapse of the Warsaw Pact. The Poles have forged special relationships with Lithuania and particularly Ukraine,[51] with whom they will co-host the UEFA Euro 2012 football tournament, in an effort to firmly anchor these countries within the Western world and provide them with an alternative to aligning themselves with the Russian Federation respectively. Despite many positive developments in the region, Poland has found itself in a position where it must seek to defend the rights of ethnic Poles living in the former Soviet Union; this is particularly true of Belarus, where in 2005 the Lukashenko regime launched a campaign against the Polish ethnic minority.[52]

Prime Minister Donald Tusk (right) arrives with former President Lech Wałęsa for the EPP party congress in Warsaw

Poland is the sixth most populous member state of the European Union and, ever since joining in 2004, has pursued policies to increase its role in European affairs. Poland has a grand total of 51 representatives in the European Parliament and in addition to this, since 14 July 2009, former Prime Minister of Poland Jerzy Buzek, has been President of the European Parliament.[53]

Administrative divisions

Poland's current voivodeships (provinces) are largely based on the country's historic regions, whereas those of the past two decades (to 1998) had been centred on and named for individual cities. The new units range in area from less than 10000 square kilometres (**unknown operator: u'strong'** sq mi) for Opole Voivodeship to more than 35000 square kilometres (**unknown operator: u'strong'** sq mi) for Masovian Voivodeship. Administrative authority at voivodeship level is shared between a government-appointed voivode (governor), an elected regional assembly (*sejmik*) and an executive elected by that assembly.

The voivodeships are subdivided into *powiats* (often referred to in English as counties), and these are further divided into *gminas* (also known as communes or municipalities). Major cities normally have the status of both *gmina* and *powiat*. Poland currently has 16 voivodeships, 379 powiats (including 65 cities with *powiat* status), and 2,478 *gminas*.

Voivodeship		Capital city or cities
in English	**in Polish**	
Greater Poland	*Wielkopolskie*	Poznań
Kuyavian-Pomeranian	*Kujawsko-Pomorskie*	Bydgoszcz / Toruń
Lesser Poland	*Małopolskie*	Kraków
Łódź	*Łódzkie*	Łódź
Lower Silesian	*Dolnośląskie*	Wrocław
Lublin	*Lubelskie*	Lublin
Lubusz	*Lubuskie*	Gorzów Wielkopolski / Zielona Góra
Masovian	*Mazowieckie*	Warsaw
Opole	*Opolskie*	Opole
Podlaskie	*Podlaskie*	Białystok
Pomeranian	*Pomorskie*	Gdańsk
Silesian	*Śląskie*	Katowice
Subcarpathian	*Podkarpackie*	Rzeszów
Świętokrzyskie (Holy Cross)	*Świętokrzyskie*	Kielce
Warmian-Masurian	*Warmińsko-Mazurskie*	Olsztyn
West Pomeranian	*Zachodniopomorskie*	Szczecin

Military

The Polish armed forces are composed of four branches: Land Forces (*Wojska Lądowe*), Navy (*Marynarka Wojenna*), Air Force (*Siły Powietrzne*) and Special Forces (*Wojska Specjalne*). The military is subordinate to the Minister for National Defence, however its sole commander in chief is the President of the Republic.

Polish Air Force F-16 Fighting Falcon

The Polish army currently consists of 65,000 active personnel, whilst the navy and air force respectively employ 14,300 and 26,126 servicemen and women. The Polish Navy is one of the larger navies on the Baltic Sea and is mostly involved in Baltic operations such as search and rescue provision for the section of the Baltic under Polish command, as well as hydrographic measurements and research; recently however, the Polish Navy played a more international role as part of the 2003 invasion of Iraq, providing logistical support for the United States Navy. The current position of the Polish Air Force is much the same; it has routinely taken part in Baltic Air Policing assignments, but otherwise, with the exception of a number of units serving in Afghanistan, has seen no active combat since the end of the Second World War. In 2003, the F-16C Block 52 was chosen as the new general multi-role fighter for the air force, the first deliveries taking place in November 2006; it is expected (2010) that the Polish Air Force will create three squadrons of F-16s, which will all be fully operational by 2012.

Polish Army vehicles and troops on patrol in Afghanistan

The most important mission of the armed forces is the defence of Polish territorial integrity and Polish interests abroad.[54] Poland's national security goal is to further integrate with NATO and European defence, economic, and political institutions through the modernisation and reorganisation of its military.[54] Currently the armed forces is being re-organised according to NATO standards, and as of 1 January 2010, the transition to an entirely contract-based military has been completed. Previously male citizens were expected to complete a period of active service with the military; since 2007 up until the amendment of the law on conscription, the obligatory term of service was nine months.[55]

Polish military doctrine reflects the same defensive nature as that of its NATO partners. From 1953 to 2009 Poland was a large contributor to various United Nations peacekeeping missions.[54] [56] The Polish Armed Forces took part in the 2003 invasion of Iraq, deploying 2,500 soldiers in the south of that country and commanding the 17-nation Multinational force in Iraq.

The military was temporarily, but severely, affected by the loss of many of its top commanders in the wake the 2010 Polish Air Force Tu-154 crash near Smolensk, Russia, which killed all 96 passengers and crew, including, amongst others, the Chief of the Polish Army's General Staff Franciszek Gągor and Polish Air Force commanding general Andrzej Błasik. They were en route from Warsaw to attend an event to mark the 70th anniversary of the Katyn massacre, whose site is commemorated approximately 19 km west of Smolensk.[57] [58]

Law enforcement and emergency services

Mounted officers of the State Police in Chorzów

Poland has a highly developed system of law enforcement with a long history of effective policing by the State Police Service. The structure of law enforcement agencies within Poland is a multi-tier one, with the State Police providing criminal-investigative services, Municipal Police serving to maintain public order and a number of other specialised agencies, such as the Polish Border Guard, acting to fulfil their assigned missions. In addition to these state services, private security companies are also common, although they possess no powers assigned to state agencies, such as, for example, the power to make an arrest or detain a suspect.

Emergency services in Poland consist of the Emergency Medical Services, Search and Rescue units of the Polish Armed Forces and State Fire Service. Emergency medical services in Poland are, unlike other services, provided for by local and regional government.

Since joining the European Union all of Poland's emergency services have been undergoing major restructuring and have, in the process, acquired large amounts of new equipment and staff.[59] All emergency services personnel are now uniformed and can be easily recognised thanks to a number of innovative design features, such as reflective paint and printing, present throughout their service dress and vehicle liveries. In addition to this, in an effort to comply with EU standards and safety regulations, the police and other agencies have been steadily replacing and modernising their fleets of vehicles; this has left them with thousands of new automobiles, as well as many new aircraft, boats and helicopters.[60]

Economy

Financial centre of Warsaw

Poland's high-income economy[61] is considered to be one of the healthiest of the post-Communist countries and is currently one of the fastest growing within the EU. Having a strong domestic market, low private debt, flexible currency, and not being dependent on a single export sector, Poland is the only European economy to have avoided the late-2000s recession.[62] Since the fall of the communist government, Poland has steadfastly pursued a policy of liberalising the economy and today stands out as a successful example of the transition from a centrally planned economy to a primarily market-based economy. In 2009 Poland had the highest GDP growth in the EU. As of February 2012, the Polish economy has not entered a recession in the wake of the global financial crisis.[63] [64]

The privatization of small and medium state-owned companies and a liberal law on establishing new firms have allowed the development of an aggressive private sector. As a consequence, consumer rights organizations have also appeared. Restructuring and privatisation of "sensitive sectors" such as coal, steel, rail transport and energy has been continuing since 1990. Between 2007 and 2010, the government plans to float twenty public companies on the Warsaw Stock Exchange, including parts of the coal industry. The biggest privatisations have been the sale of the national telecoms firm Telekomunikacja Polska to France Télécom in 2000, and an issue of 30% of the shares in Poland's largest bank, PKO Bank Polski, on the Polish stockmarket in 2004.

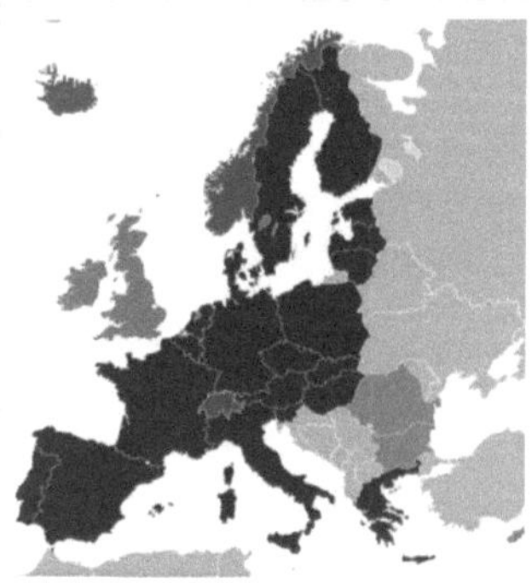

Poland is part of the Schengen Area and the EU single market.

The Polish banking sector is the largest in central and eastern Europe as well being as the largest and the most highly developed sector of the country's financial markets. It is regulated by the Polish Financial Supervision Authority. During the transformation to a market-oriented economy, the government privatized some banks, recapitalized the rest and introduced legal reforms that made the sector competitive. This has attracted a significant number of strategic foreign investors. Poland's banking sector has approximately 5 domestic banks, a network of nearly 600 cooperative banks and 18 branches of foreign-owned banks. In addition, foreign investors have controlling stakes in nearly 40 commercial banks, which make up 68% of the banking capital.[65]

Poland has a large number of private farms in its agricultural sector, with the potential to become a leading producer of food in the European Union. Structural reforms in health care, education, the pension system, and state administration have resulted in larger-than-expected fiscal pressures. Warsaw leads Central Europe in foreign investment.[66] GDP growth had been strong and steady from 1993 to 2000 with only a short slowdown from 2001 to 2002.

The economy had growth of 3.7% annually in 2003, a rise from 1.4% annually in 2002. In 2004, GDP growth equaled 5.4%, in 2005 3.3% and in 2006 6.2%.[67] According to Eurostat data, Polish PPS GDP per capita stood at 61% of the EU average in 2009.[68]

Gdynia, situated at Gdańsk Bay on the south coast of the Baltic Sea, is an important seaport of Poland.

Although the Polish economy is currently undergoing economic development, there are many challenges ahead. The most notable task on the horizon is the preparation of the economy (through continuing deep structural reforms) to allow Poland to meet the strict economic criteria for entry into the Eurozone. According to the Polish foreign minister Radosław Sikorski the country could join the eurozone before 2016.[69] Some businesses may already accept the euro as payment. In addition, the ability to establish and conduct business easily has been cause for economic hardship as the World Economic Forum recently ranked Poland near the bottom of OECD countries in terms of the clarity, efficiency and neutrality of its legal framework for firm to settle disputes.[70] A report concluded that on-going foreign business disputes issues may "have damaged Poland's reputation as an attractive location for FDI" by reinforcing the impression of "Poland's substandard reputation for maintaining an efficient and neutral framework to settle business disputes involving multinational foreign investors."[71] Ernst & Young's 2010 European attractiveness survey reported that Poland saw a 52% decrease in FDI job creation and a 42% decrease in number of FDI projects since 2008.[72]

Average salaries in the enterprise sector in December 2010 were 3,848 PLN (1,012 euro or 1,374 US dollars)[73] and growing sharply.[74] Salaries vary between the regions: the median wage in the capital city Warsaw was 4,603 PLN (1,177 euro or 1,680 US dollars) while in Kielce it was only 3,083 PLN (788 euro or 1125 US dollars). Differences in salaries in various districts of Poland is even higher and range from 2,020 PLN (517 euro or 737 US dollars) in Kępno County, which is located in Greater Poland Voivodeship to 5,616 (1,436 euro or 2,050 US dollars) in Lubin County, which lies in Lower Silesian Voivodeship.[75]

According to a Credit Suisse report, Poles are the second wealthiest (after Czechs) of the Central European peoples.[76] [77] This makes Poland an attractive destination for many guest workers particularly from Ukraine, Belarus, Russia and Vietnam.[78] [79] [80] Even though Poland is rather an ethnically homogeneous country, the number of foreigners is growing every year.[80] [81]

Since the United Kingdom, Ireland and some other European countries opened their job markets for Poles, many workers, especially from rural regions, have left the country to seek a better wages abroad. However, there is a rapid growth of the salaries, booming economy, strong value of Polish currency, and quickly decreasing unemployment (from 14.2% in May 2006 to 6.7% in August 2008).[82] Commodities produced in Poland include: electronics, cars (Arrinera, Leopard), buses (Autosan, Solaris, Solbus), helicopters (PZL Świdnik), transport equipment, locomotives, planes (PZL Mielec), ships, military engineering (including tanks, SPAAG systems), medicines (Polpharma, Polfa), food, clothes, glass, pottery (Bolesławiec), chemical products and others.

Corporations

Poland is recognised as a regional economic power within Central Europe, possessing nearly 40 percent of the 500 biggest companies in the region (by revenues).[83] Poland was the only member of the EU to avoid the recession of the late 2000s, a testament to the Polish economy's stability.[62] The country's most competitive firms are components of the WIG20 which is traded on the Warsaw Stock Exchange.

The Warsaw Stock Exchange is, by market capitalisation, one of Central Europe's largest

Well known Polish brands include, amongst others, PKO BP, PKN Orlen, PGE, PZU, PGNiG, Tauron Group, Lotos Group, KGHM Polska Miedź, Telekomunikacja Polska, Plus, Play, PLL LOT, Poczta Polska, PKP, Biedronka, BRE Bank, Getin Holding and TVP.[84]

Poland is recognised as having an economy with significant development potential, overtaking the Netherlands in mid-2010 to become Europe's sixth largest economy.[85] Foreign Direct Investment in Poland has remained strong ever since the country's re-democratisation following the Round Table Agreement in 1989. Despite this, problems do exist, and further progress in achieving success depends largely on the government's privatisation of Poland's remaining state industries and continuing development and modernisation of the economy.

Warsaw is home to many of Poland's largest business enterprises

The list includes the largest companies by turnover in 2009, but does not include major banks or insurance companies:

Rank in 2009[86]	Name of concern	Location of headquarters	Revenue (Thou. PLN)	Profit (Thou. PLN)	Employees
1.	PKN Orlen SA	Płock	49,876,074	1,907,812	4,482
2.	PGE SA	Lublin	22,418,183	5,378,534	46,357
3.	Fiat Auto Poland SA	Bielsko-Biała	19,788,511	820,420	6,421
4.	PGNiG SA	Warsaw	19,493,756	1,442,103	31,685
5.	PBG SA	Poznań	17,761,936	*N/A*	27,032
6.	Metro Group Poland	Warsaw	16,800,000	*N/A*	24,077
7.	Telekomunikacja Polska SA	Warsaw	16,560,000	1,597,000	27,667
8.	Lotos Group SA	Gdańsk	15,379,754	1,091,950	4,949
9.	Tauron Group SA	Katowice	13,859,135	1,165,020	28,839
10.	KGHM Polska Miedź SA	Lubin	12,286,107	3,070,799	18,370

Tourism

Poland is a major part of the global tourism market and is currently experiencing an upward trend in its number of visitors; this began shortly after joining the European Union.[87] Tourism in Poland contributes to the country's overall economy and makes up a relatively large proportion of the country's service market.

The city of Zamość is a UNESCO World Heritage Site

The most attractive urban destinations for tourists are:

- Kraków - The former Polish capital, headquarters and the place of coronation of Polish kings.
- Wrocław - One of the oldest cities in Poland with a very checkered history over the centuries included in the many kingdoms and countries. A huge number of monuments, great architecture, lots of parks. European Capital of Culture 2016.
- Gdańsk
- Warsaw - present Polish capital
- Poznań
- Lublin
- Toruń

Centennial Hall in Wrocław is a UNESCO World Heritage Site

in addition to these the historic site of the Auschwitz German concentration camp near Oświęcim is a noteworthy place of pilgrimage and a now constitutes a major monument to the prevention of war and suffering.

Popular areas of natural beauty include north Poland's Baltic Sea coast, Masurian Lake District and Białowieża Forest, in southern Karkonosze, Table Mountains, Tatra Mountains and Bieszczady Mountains.

Poland's main tourist offerings are thought to be based around city-sightseeing and extra-urban historical monuments, business trips, qualified tourism, agrotourism, mountain hiking and climbing among others.

Poland was the 17th most visited country by foreign tourists in 2008.[88]

Energy

Żerań power station in Warsaw

The electricity generation sector in Poland is still largely fossil-fuel based. Many power plants nationwide use Poland's position as a major European exporter of coal to their advantage by continuing to use coal as the primary raw material in production of their energy; in 2007, hard bituminous coal contributed 48% of energy generation, brown coal and gas 12% each and oil 23%.[89] Currently the three largest Polish coal mining firms (*Weglokoks, Kompania Węglowa and JSW*) extract around 100 million tonnes of coal annually; all three of these companies are key constituents of the Warsaw Stock Exchange's lead economic indexes.

Renewable forms of energy currently only account for a small proportion of Poland's full energy generation capacity.[90] However, the national government has set targets for the development of renewable energy sources in Poland which should see the portion of power produced by renewable resources climb to 7.5% by 2010 and 15% by 2020. This is to be achieved mainly through the construction of wind farms and a number of hydroelectric stations.

Poland is thought to have around 164,800,000,000 m³ of proven natural gas reserves and around 96,380,000 barrels of proven oil reserves. These reserves are currently attended to and exploited by energy supply companies such as PKN Orlen (*the only Polish company listed in the Fortune Global 500*). However, due to the small amounts of fossil fuels naturally occurring in Poland not being enough to satisfy the full energy consumption needs of the population and thus need to buy from abroad, the country is considered to be a net importer of oil and natural gas.

Transport

Today transport in Poland is provided by means of rail, road, shipping and air travel. Positioned in Central Europe and with an eastern border compromising the largest external border of the Schengen Area with the rest of East-Central Europe, Poland has long been, and remains a key country through which imports to and exports from the European Union pass.

LOT was the first European airline to introduce the Boeing 787 Dreamliner

Since joining the EU in 2004, Poland has invested large amounts of money into the modernisation of its transport networks. The country now has a developing expressways network composed of motorways such as the A1, A2, A4, A18 and express roads such as the S7, S8. In addition to these newly built roads, many local and regional roads are being rebuilt as part of a national programme to rebuild all roads in Poland.[91]

The A4 Motorway near Kraków

Again, with regard to railways, much the same situation is taking place. The Polish authorities have begun a program by which they hope to increase operating speeds across the entire Polish rail network; this is particularly true of a number of national trunk routes which are expected to soon receive new rolling stock capable of speeds over 200 km/h. Finally, there is a plan to introduce high speed rail to Poland from around 2014. The Polish government recently revealed that it intends to connect all major cities to a future high-speed rail network by 2020.[92] Most intercity rail operations in Poland are operated by PKP Intercity whilst regional trains are run by a number of operators, the largest of which is Przewozy Regionalne.

The air and maritime transport markets in Poland are largely well developed. Poland has a number of international airports; the largest of which is Warsaw Chopin Airport, the primary global hub for LOT Polish Airlines, which is the largest airline of Central and Eastern Europe and one of the world's oldest airlines still in operation today. Seaports exist all along Poland's Baltic coast, with most freight operations using either Gdynia or Gdańsk as their base. Passenger ferries link Poland with Scandinavia all year round; these services are provided from Gdańsk by Polferries, Stena Line from Gdynia and Unity Line from the Port of Świnoujście.

Science

In 1925, the chemist Maria Skłodowska-Curie established the first Radium Institute in Poland.[93]

According to Frost & Sullivan's Country Industry Forecast the country is becoming an interesting location for research and development investments.[94] Multinational companies such as: ABB, Delphi, GlaxoSmithKline, Google, Hewlett–Packard, IBM, Intel, LG Electronics, Microsoft, Motorola, Siemens and Samsung have set up research and development centres in Poland.[95] Over 40 research and development centers and 4,500 researchers make Poland the biggest research and development hub in Central and Eastern Europe.[79] [94] Companies chose Poland because of the availability of highly qualified labour force, presence of universities, support of authorities, and the largest market in Central Europe.[94]

Warsaw's Staszic Palace is home to the Polish Academy of Sciences

Today Poland's tertiary education institutions; traditional universities (found in its major cities), as well as technical, medical, and economic institutions, employ around 61,000 researchers and members of staff. There are around 300 research and development institutes, with about 10,000 researchers. In total, there are around 91,000 scientists in Poland today. However, in the 19th and 20th centuries many Polish scientists worked abroad; one of the greatest of these exiles was Maria Skłodowska-Curie, a physicist and chemist who lived much of her life in France. In the first half of the 20th century, Poland was a flourishing centre of mathematics. Outstanding Polish mathematicians formed the Lwów School of Mathematics (with Stefan Banach, Hugo Steinhaus, Stanisław Ulam) and Warsaw School of Mathematics (with Alfred Tarski, Kazimierz Kuratowski, Wacław Sierpiński). The events of World War II pushed many of them into exile. Such was the case of Benoît Mandelbrot, whose family left Poland when he was still a child. An alumnus of the Warsaw School of Mathematics was Antoni Zygmund, one of the shapers of 20th-century mathematical analysis.

According to a KPMG report[96] 80% of Poland's current investors are content with their choice and willing to reinvest. In 2006, Intel decided to double the number of employees in its research and development centre in Gdańsk.[95]

Communications

The Main Municipal Post Office of Bydgoszcz

The share of the telecom sector in the GDP is 4.4% (end of 2000 figure), compared to 2.5% in 1996. Nevertheless, despite high expenditures for telecom infrastructure (the coverage increased from 78 users per 1,000 inhabitants in 1989 to 282 in 2000).

The value of the telecommunication market is zl 38.2bn (2006), and it grew by 12.4% in 2007 PMR.[97] The coverage mobile cellular is over 1000 users per 1000 people (2007). Telephones—mobile cellular: 38.7 million (Onet.pl & GUS Report, 2007), telephones—main lines in use: 12.5 million (Telecom Team Report, 2005).

With regard to internet access, the most popular ADSL services for home users in Poland are Neostrada provided by TPSA, and Net24 provided by Netia. Business users as well as some home users use Internet DSL TP also offered by TPSA. According to Eurostat, OECD and others, Internet access in Poland is amidst the most expensive in Europe. This is mostly caused by the lack of competitiveness. New operators, such as Dialog and GTS Energis are making their own provider lines and offer more attractive and cheaper service. Recently, the Polish Office of Electronical Communication passed a bill forcing the TPSA to rent 51% of their ADSL lines to other ISPs for 60% lower prices. This move will definitely affect the prices of DSL in Poland.

TP S.A. headquarters in Warsaw

The public postal service in Poland is operated by Poczta Polska (The Polish Post). It was created on October 18, 1558, when king Zygmunt August established a permanent postal route from Kraków to Venice (later also to Wilno) in order to manage affairs in Italy that arose after the death of Queen Bona, his mother. Since then the service was dissolved on a number of occasions, most notably during the partitions of Poland. After regaining independence in 1918, the united territory of Poland was in need of a uniform network of communication. Thus, the interwar period saw the rapid development of the postal system as new services were introduced (e.g. money transfers, payment of pensions, delivery of magazines, air mail). Although during national uprisings and in the course of wars communication was provided mainly through field post, which was subject to military authority, postmen always took active part in the fight for independence by secretly delivering parcels and documents, or by providing vital information about the enemy. Many important events in the history of Poland involved the postal service, like the heroic Defence of the Polish Post Office in Gdańsk in 1939 and the participation of the Polish Scouts' Postal Service in the Warsaw Uprising. During the difficult times of the Second World War, the Polish Post in exile would lift up the spirits of compatriots by issuing postage stamps. Nowadays the service is a modern, functioning state-owned company which provides a number of standard and express delivery options, as well as operating the Polish postal home-delivery service. The postal service is currently expanding into the provision of logistical services.

Demographics

Poland, with 38,116,000 inhabitants,[5] has the eighth-largest population in Europe and the sixth-largest in the European Union. It has a population density of 122 inhabitants per square kilometer (328 per square mile).

Long Market in Gdańsk filled with picturesque Dutch style tenements is a favourite meeting place in the Kashubian capital.

Poland historically contained many languages, cultures and religions on its soil. The country had a particularly large Jewish population prior to World War II, when the Nazi Germany's regime led to The Holocaust. It caused Poland's Jewish population, estimated at 3 million before the war, to drop to just 300,000. The outcome of the war, particularly the shift of Poland's borders to the area between the Curzon Line and the Oder-Neisse line, coupled with post-war expulsion of minorities, significantly reduced the country's ethnic diversity. Over 7 million Germans fled or were expelled from the Polish side of the Oder-Neisse boundary.[98]

According to the 2002 census, 36,983,700 people, or 96.74% of the population, consider themselves Polish, while 471,500 (1.23%) declared another nationality, and 774,900 (2.03%) did not declare any nationality. The largest minority nationalities and ethnic groups in Poland are Silesians (173,153 according to the census), Germans (152,897 according to the census, 92% in Opole Voivodeship and Silesian Voivodeship), Belarusians (c. 49,000), Ukrainians (c. 30,000), Lithuanians, Russians, Roma, Jews, Lemkos, Slovaks, Czechs, and Lipka Tatars.[99] Among foreign citizens, the Vietnamese are the largest ethnic group, followed by Greeks and Armenians.

The Polish language, part of the West Slavic branch of the Slavic languages, functions as the official language of Poland. Until recent decades Russian was commonly learned as a second language but has been replaced by English and German as the most common second languages studied and spoken.[100]

The Main Market Square in Kraków is the heart of Poland's southern cultural capital

In recent years, Poland's population has decreased because of an increase in emigration and a sharp drop in the birth rate. Since Poland's accession to the European Union, a significant number of Poles have emigrated, primarily to the United Kingdom, Germany and Ireland in search of work and better work opportunities abroad. In April 2007, the Polish population of the United Kingdom had risen to approximately 300,000, and estimates place the Polish population in Ireland at 65,000. Some sources claim that the number of Polish citizens who emigrated to the UK after 2004 is as high as 2 million.[101] This, however, is contrasted by a recent trend that shows that more Poles are entering the country than leaving it.[102]

Polish minorities are still present in the neighboring countries of Ukraine, Belarus, and Lithuania, as well as in other countries (see Poles for population numbers). Altogether, the number of ethnic Poles living abroad is estimated to be around 20 million.[103] The largest number of Poles outside of Poland can be found in the United States.[104]

Religion

Basilica of Our Lady of Licheń

Until World War II Poland and Prussia had a Protestant majority, although a religiously diverse society, in which substantial Jewish, Christian Orthodox and Roman Catholic minorities coexisted. As a result of the Holocaust and the post–World War II flight and expulsion of German and Ukrainian populations, Poland has become overwhelmingly Roman Catholic. In 2007, 88.4% of the population belonged to the Catholic Church.[105] Though rates of religious observance are lower, at 52%[106] or 51% of the Polish Catholics,[107] Poland remains one of the most devoutly religious countries in Europe.[108]

Holy Spirit Orthodox Church in Białystok

From 16 October 1978 until his death on 2 April 2005 Karol Józef Wojtyła (later Pope John Paul II), a natural born Pole, reigned as Supreme Pontiff of the Roman Catholic Church and Sovereign of Vatican City. His was the second-longest documented pontificate; only Pope Pius IX served longer. He has been the only Slavic and Polish Pope to date, and was the first non-Italian Pope since Dutch Pope Adrian VI in 1522.[109] Additionally he is credited with having played a significant role in hastening the downfall of communism in Poland and throughout Central and Eastern Europe; he is famously quoted as having, at the height of communism in 1979, told Poles "not be afraid", later praying: "Let your Spirit descend and change the image of the land... this land".[110] [111] He is a deeply revered figure within Polish society, and his passing in 2005 was met with large-scale outpourings of national grief.

Kraków's Tempel Synagogue is one of the largest in Poland still functioning today.

Religious minorities include Polish Orthodox (about 506,800),[5] various Protestants (about 150,000),[5] Jehovah's Witnesses (126,827),[5] Eastern Catholics, Mariavites, Polish Catholics, Jews, and Muslims (including the Tatars of Białystok). Members of Protestant churches include about 77,500 in the largest Evangelical-Augsburg Church,[5] and a similar number in smaller Pentecostal and Evangelical churches.

Freedom of religion is now guaranteed by the 1989 statute of the Polish Constitution,[112] enabling the emergence of additional denominations.[113] However, because of pressure from the Polish Episcopate, the exposition of doctrine has entered the public education system as well.[114] [115] According to a 2007 survey, 72% of respondents were not opposed to religious instruction in public schools; alternative courses in ethics are available only in one percent of the entire public educational system.[116]

Famous sites of Christian pilgrimage in Poland include the Monastery of Jasna Góra in the southern Polish city of Częstochowa, as well as the Family home of John Paul II in Wadowice just outside of Kraków.

Health

Poland's healthcare system is based on an all-inclusive insurance system. State subsidised healthcare is available to all Polish citizens who are covered by this general health insurance program. However, it is not compulsory to be treated in a state-run hospital as a number of private medical complexes do exist nationwide.[117]

All medical service providers and hospitals in Poland are subordinate to the Polish Ministry of Health, which provides oversight and scrutiny of general medical practice as well as being responsible for the day to day administration of the healthcare system. In addition to these roles, the ministry is also tasked with the maintenance of standards of hygiene and patient-care.

Main clinical building of the Gdańsk University Medical Centre in Wrzeszcz

Hospitals in Poland are organised according to the regional administrative structure, resultantly most towns have their own hospital *(Szpital Miejski)*. Larger and more specialised medical complexes tend only to be found in major cities, with some even more specialised units located only in the capital, Warsaw. However, all voivodeships have their own general hospital (most have more than one), all of which are obliged to have a trauma centre; these types of hospital, which are able to deal with almost all medical problems are called 'regional hospitals' *(Szpital Wojewódzki)*. The last category of hospital in Poland is that of specialised medical centres, an example of which would be the Skłodowska-Curie Institute of Oncology, Poland's leading, and most highly specialised centre for the research and treatment of cancer.

The Polish health-care industry is currently undergoing a major transformation, with many hospitals being listed as top priorities for refurbishment.[118] As a result of this process, many hospitals have already been thoroughly modernised throughout and are now equipped with the latest in medical hardware. The overall quality of healthcare provision nationwide, as judged by European standards, is generally regarded as being very high.[119] This is reflected in the nation's average life expectancy, which at 71 for males and 80 for females,[120] has shown a marked increase from 63/68 in 2003, and now corresponds with the average figures for life expectancy in the European Union.

Education

The education of Polish society was a goal of rulers as early as the 12th century, and Poland soon became one of the most educated countries in Europe. The library catalogue of the Cathedral Chapter of Kraków dating back to 1110 shows that in the early 12th-century Polish intellectuals had access to European literature. The Jagiellonian University, founded in 1364 by King Casimir III in Kraków, is one of Europe's oldest universities. In 1773 King Stanisław August Poniatowski established the Commission of National Education *(Komisja Edukacji Narodowej)*, the world's first state ministry of education.

The *Collegium Maius* is the oldest building of the Jagiellonian University in Kraków

The Adam Mickiewicz University, Poznań

The first university in Poland, Kraków's Jagiellonian University, was established in 1364 by Casimir III the Great in Kraków. It is the oldest university in Poland. It is the second oldest university in Central Europe and one of the oldest universities in the world. Casimir III realized that the nation needed a class of educated people, especially lawyers, who could codify the country's laws and administer the courts and offices. His efforts to found an institution of higher learning in Poland were finally rewarded when Pope Urban V granted him permission to open the University of Krakow.

Since changes made in 2009 education in Poland starts at the age of five or six for the 0 class (Kindergarten) and six or seven years in the 1st class of primary school (Polish *szkoła podstawowa*). It is compulsory that children do one year of formal education before entering 1st class at no later than 7 years of age.

The wearing of traditional academic dress is an important feature of Polish educational ceremonies

At the end of 6th class when the students are 13, they take a compulsory exam that will determine to which lower secondary school (*gimnazjum, pronounced gheem-nah-sium*) (Middle School/Junior High) they will be accepted. They will attend this school for three years for classes, 7, 8, and 9. They then take another compulsory exam to determine the upper secondary level school they will attend. There are several alternatives, the most common being the three years in a *liceum* or four years in a technikum. Both end with a maturity examination (matura, quite similar to French baccalauréat), and may be followed by several forms of upper education, leading to licencjat or inżynier (the Polish Bologna Process first cycle qualification), magister (second cycle qualification) and eventually doktor (third cycle qualification).[121]

The University of Wrocław at night.

There are currently 18 fully accredited traditional universities in Poland, these are then further supplemented by 20 technical universities, nine independent medical universities and five universities for the study of economics. In addition to these institutions there are then nine agricultural academies, three pedagogical universities, a theological academy and three maritime service universities. Poland's long history of promoting the arts has led to the establishment of a number of higher educational institutes dedicated to the teaching of the arts. Amongst these are the seven higher state academies of music. All of these institutions are further supplemented by a large number of private educational institutions and the four national military academies (two for the army and one for each of the other branches of service), bringing the total number of organisations for the pursuit of higher education to well over 500, one of the largest numbers in Europe. The Programme for International Student Assessment, coordinated by the OECD, currently ranks Poland's educational system as the 23rd best in the world, being neither significantly higher nor lower than the OECD average.[122]

Culture

The culture of Poland is closely connected with its intricate 1000 year history[123] Its unique character developed as a result of its geography at the confluence of European cultures. With origins in the culture of the Proto-Slavs, over time Polish culture has been profoundly influenced by its interweaving ties with the Germanic, Latinate and Byzantine worlds as well as in continual dialog with the many other ethnic groups and minorities living in Poland.[124] The people of Poland have traditionally been seen as hospitable to artists from abroad and eager to follow cultural and artistic trends popular in other countries. In the 19th and 20th centuries the Polish focus on cultural advancement often took precedence over political and economic activity. These factors have contributed to the versatile nature of Polish art, with all its complex nuances.[124]

Famous people

Mikołaj Kopernik
(Latin: *Nicolaus Copernicus*)

Poland is the birthplace of many distinguished personalities (see. List of Polish people), among which are: Mikołaj Kopernik,[125] Fryderyk Chopin,[126] [127] Maria Skłodowska Curie,[128] Stanislaw Ulam, Ernest Malinowski, Tadeusz Kościuszko, Kazimierz Pułaski, Józef Piłsudski, Lech Wałęsa and Pope John Paul II (Karol Wojtyła). Great Polish painter Jan Matejko devoted his monumental art to the most significant historical events on Polish lands, along with the playwright, painter and poet Stanisław Wyspiański. Stanisław Ignacy Witkiewicz (Witkacy) was an example of a Polish avant-garde philosopher and author of aesthetic theories.

Polish literature dates back to the 12th century[129] and includes many famous poets and writers such as Jan Kochanowski, Adam Mickiewicz, Bolesław Prus, Juliusz Słowacki, Witold Gombrowicz, Stanisław Lem and, Ryszard Kapuściński. Writers Henryk Sienkiewicz, Władysław Reymont, Czesław Miłosz, Wisława Szymborska have each won the Nobel Prize in Literature. Also a renowned Polish novelist, who wrote in the English language, was Joseph Conrad.[130]

Many world famous Polish movie directors include Academy Awards winners Roman Polański, Andrzej Wajda, Zbigniew Rybczyński, Janusz Kamiński, Krzysztof Kieślowski, Agnieszka Holland. World renowned actresses were Helena Modjeska and Pola Negri.

Poles have outstanding achievements in mountaineering in the Himalayas, especially in the winter gathering Eight-thousander. The most famous Polish climbers are Jerzy Kukuczka, Krzysztof Wielicki, Piotr Pustelnik, Andrzej Zawada, Maciej Berbeka and Wanda Rutkiewicz.

Society

The Grand Hotel in Sopot. Poland's Baltic Sea resorts are very popular tourist destinations amongst Poles as well as foreign tourists

Pope John Paul II is considered to have been a great promoter of Poland around the world

Poland has a great, long standing tradition of tolerance towards minorities, as well as absence of discrimination on the grounds of religion, nationality or race. This is for historical reasons and that is ethnic minorities in Poland are very few. It has a high level of gender equality, promotes disability rights movement and promotes equality. Poland is the first country in the world where corporal punishment was prohibited.[131] Poland has, throughout most of its long history, experienced only very limited immigration from abroad; this trend can largely be attributed to Poland's lack of slavery and overseas colonies as well as its lack of existence as a state during much of the 19th and early 20th centuries. Despite this, the country has for a long time been regarded as having a very tolerant society, which affords equal rights to all people no matter what their ethnic background. This can be said to stem largely from the reign of King Casimir III the Great and his acceptance for Poland's Jewish community, in a time when the most of Europe recessed antisemitic mood. The history of Jews in Poland shows peaceful co-existence of a nation and particular ethnic group.

As many as 96.7% of Polish citizens declare to be Poles, and 97.8% declare that they speak Polish at home (Census 2002). The population of Poland became one of the most ethnically homogeneous in the world as a result of the radically altered borders after World War II and the subsequent migrations. This homogeneity is a result of post World War II deportations ordered by the Soviet authorities, who wished to remove the sizeable Polish minorities from Lithuania, Belarus and Ukraine and repatriation of Ukrainians from Poland to the Soviet Union (see territorial changes of Poland and historical demography of Poland for details). Unlike in many other countries, the ethnic minority rights in Poland are guaranteed directly in the Constitution of Poland (art. 35), and today there are, amongst others, sizeable German, Ukrainian and Belarusian minorities present in the country.[132]

After the formal collapse of Communism in 1989, Poland greatly improved its image in the world and thus has received further support from the country's recent economic success and effective entry into the structures of the European Union. Polish citizens have obtained a good reputation as workers in the united Europe, mainly due to the broad range of jobs beyond the borders of their state, since 2004. The results of an Organization for Security and Co-operation in Europe (OSCE) survey from 2004 showed that Poles work the second most hours per week of any nationality worldwide.

Poland remains one of the most peaceful countries in the world.[133]

Music

Fryderyk Chopin, composer

Artists from Poland, including famous composers like Chopin or Penderecki and traditional, regionalized folk musicians, create a lively and diverse music scene, which even recognizes its own music genres, such as poezja śpiewana and disco polo. As of 2006, Poland is one of the few countries in Europe where rock and hip hop dominate over pop music, while all kinds of alternative music genres are encouraged.

The origins of Polish music can be traced as far back as the 13th century; manuscripts have been found in Stary Sącz, containing polyphonic compositions related to the Parisian Notre Dame School. Other early compositions, such as the melody of *Bogurodzica* and *Bóg się rodzi* (a coronation polonaise for Polish kings by an unknown composer), may also date back to this period, however, the first known notable composer, Mikołaj z Radomia, was born and lived in the 15th century. During the 16th century, two main musical groups – both based in Kraków and belonging to the King and Archbishop of the Wawel – led to the rapid development of Polish music. Composers writing during this period include Wacław z Szamotuł, Mikołaj Zieleński, and Mikołaj Gomółka. Diomedes Cato, a native-born Italian who lived in Kraków from about the age of five, became one of the most famous lutenists at the court of Sigismund III, and not only imported some of the musical styles from southern Europe, but blended them with native folk music.[134]

Tomasz Stańko is a popular contemporary Polish jazz musician

At the end of the 18th century, Polish classical music evolved into national forms like the polonaise. In the 19th century the most popular composers were: Józef Elsner and his pupils Fryderyk Chopin and Ignacy Dobrzyński. Important opera composers of the era were Karol Kurpiński and Stanisław Moniuszko whilst the list of famous soloists and composers included Henryk Wieniawski, Juliusz Zarębski. At the turn of the 19th and 20th centuries the most promiment composers could said to have been Władysław Zeleński and Mieczysław Karłowicz, with Karol Szymanowski gaining prominence prior to World War II. Alexandre Tansman lived in Paris but had strong connections with Poland. Witold Lutosławski, Henryk Górecki, and Krzysztof Penderecki composed in Poland, Andrzej Panufnik emigrated.

Traditional Polish folk music has had a major effect on the works of many well-known Polish composers, and no more so than on Fryderyk Chopin, a widely recognised national hero of the arts. All of Chopin's works involve the piano and are technically demanding, emphasising nuance and expressive depth. As a great composer, Chopin invented the musical form known as the instrumental ballade and made major innovations to the piano sonata, mazurka, waltz, nocturne, polonaise, étude, impromptu and prélude, he was also the composer of a number of polonaises which borrowed heavily from traditional Polish folk music. It is largely thanks to him that the such pieces gained great popularity throughout Europe during the 19th century. Nowadays the most distinctive folk music can be heard in the towns and villages of the mountainous south, particularly in the region surrounding the winter resort town of Zakopane.

Today Poland has a very active music scene, with the jazz and metal genres being particularly popular amongst the contemporary populace. Polish jazz musicians such as Krzysztof Komeda, created a unique style, which was most famous in 1960s and 1970s and continues to be popular to this day. Since the fall of Communism, Poland has become a major venue for large-scale music festivals, chief among which are the Open'er Festival, Opole Festival and Sopot Festival.

Media

Headquarters of TVP in Warsaw

Poland has instituted freedom of press since the fall of communism, a system under which the media was heavily politically controlled and censored. However, public TV and radio are still regulated by the government, this is exercised through an agency called *Krajowa Rada Radiofonii i Telewizji* (*The National Radio and Television Committee*), which is similar to television regulatory commissions in other developed nations.

Poland has a number of major media outlets, chief amongst which are the national television channels. TVP is Poland's public broadcasting corporation; about a third of its income comes from a broadcast receiver licence, while the rest is made through revenue from commercials and sponsorships. State television operates two mainstream channels, TVP 1 and TVP 2, as well as regional programs (TVP Info) for each of the country's 16 voivodeships. In addition to these general channels, TVP runs a number of genre-specific programmes such as TVP Sport, TVP Historia, TVP Kultura and TVP Seriale; there are currently plans to run channels dedicated to the coverage of political affairs (TVP Parlament) and entertainment (TVP Rozrywka).

Studio set of TVP Wrocław's 'Fakty' news program

Poland has a number of internationally broadcast and 24 hour news channels, chief amongst which are Polsat News, TVN 24, and TV Polonia, the latter is a state-run channel dedicated to the transmission of Polish language television for the Polish diaspora abroad. There are a number of major private television outlets such as Polsat and the TVN network.

Poland has a highly developed printed news industry, with daily newspapers like Gazeta Wyborcza *(The Electoral Gazette)*, Rzeczpospolita *(The Republic)* and Gazeta Polska Codziennie providing more traditional, intellectually stimulating reporting and tabloids such as Fakt providing more sensationalist writing which is less current affairs orientated. Rzeczpospolita is one of the nation's oldest publications still in operation today, founded in 1920, it has become a stalwart bastion of Polish reporting and in 2006 won a prestigious award for being, along with the Guardian (a British daily), the best designed newspaper in the world.[135]

Major media outlets are experiencing an ongoing restructuring which is seeing many of them amalgamated into major media groups; a prime example of which is the German Axel Springer AG Publishing conglomerate's purchase of Fakt. International cooperation is also a growing trend within Polish media; TVP recently began cooperating with the French-German TV network ARTE.

Literature

Polish literature has a long and complicated history. During the Middle Ages most Polish authors and academics (Jan Długosz) wrote only in Latin, as at the time, this was the 'academic' language which linked Europe together; Jan Kochanowski broke this trend and became the first author to write the majority of his works in the Polish language. A number of Polish authors have won great renown in the past few centuries, however, this largely stems from the initial success of the works of Adam Mickiewicz, who wrote the first Polish epic, Pan Tadeusz, in 1834. Influential authors of the late 19th and 20th centuries include Henryk Sienkiewicz, Władysław Reymont, Witold Gombrowicz and Czesław Miłosz. To date four Polish authors have won the Nobel Prize in Literature, with Władysław Reymont being one of only nine writers to receive the prestigious award for one particular, outstanding literary work *(awarded for the great national epic, 'The Peasants' in 1924)* rather than their career as a whole.[136]

Adam Mickiewicz (1798–1855)	Henryk Sienkiewicz (1846–1916)	Władysław Reymont (1865–1925)	Witold Gombrowicz (1904–1969)	Czesław Miłosz (1911–2004)

With regard to poetry, Poland has a long and distinguished history of producing world-class poets. Chief among these are the 'three bards' (trzej wieszcze), Mickiewicz, Krasiński and Słowacki; the three national poets of Polish Romantic literature. Incidentally, the Polish word *Wieszcz* means 'prophet' or 'soothsayer', a fitting reference for the three, as the bards were thought to not only voice Polish national sentiments but to frequently foresee the nation's future.

Today the traditions of Polish literature and poetry are being carried forward by a new generation of writers. Included within this group is Wisława Szymborska, a best-selling author and recipient of the 1996 Nobel Prize in Literature.[137]

Architecture

Renaissance City Hall in Poznań

Polish cities and towns reflect the whole spectrum of European styles. Romanesque architecture is represented by St. Andrew's Church in Kraków, and characteristic for Poland Brick Gothic by St. Mary's Church in Gdańsk. Richly decorated attics and arcade loggias are the common elements of the Polish Renaissance architecture,[138] [139] like in City Hall in Poznań. For some time the late renaissance, so called mannerism, most notably in Bishop's Palace in Kielce, coexisted with the early baroque like in Church of SS. Peter and Paul in Kraków.

The Gothic St. Mary's Basilica on the Main Market Square in Kraków

History has not been kind to Poland's architectural monuments. Nonetheless, a number of ancient structures had survived: castles, churches, and stately buildings, often unique in the regional or European context. Some of them have been painstakingly restored, like Wawel Castle, or completely reconstructed after being destroyed in the Second World War, including the Old Town and Royal Castle in Warsaw and the Old Town of Gdańsk. The architecture of Gdańsk is mostly of the Hanseatic variety, a Gothic style common amongst the former trading cities along the Baltic sea and in the northern part of Central Europe. The architectural style of Wrocław is mainly representative of German architecture, since it was for centuries located within the German states. The centre of Kazimierz Dolny on the Vistula is a good example of a well-preserved medieval town. Poland's ancient capital, Kraków, ranks among the best-preserved Gothic and Renaissance urban complexes in Europe. Meanwhile, the legacy of the Kresy Marchlands of Poland's eastern regions, where Wilno and Lwów (now *Vilnius* and *Lviv*) were recognised as two major centres for the arts, played a special role in the development of Polish architecture, with Catholic church architecture deserving special note.[124]

The second half of the 17th century is marked by baroque architecture. Side towers, visible in Branicki Palace in Białystok are typical for Polish baroque. The classical Silesian baroque is represented by the University in Wrocław. Profuse decorations of Branicki Palace in Warsaw are characteristic of rococo style. The centre of Polish classicism was Warsaw under the rule of the last Polish king Stanisław August Poniatowski.[140] The Palace on the Water is the most notable example of Polish neoclassical architecture. Lublin Castle represents the Gothic Revival style in architecture, while the Izrael Poznański Palace in Łódź is an example of eclecticism.

Cuisine

Kotlet schabowy served with potatoes.

Polish cuisine has influenced the cuisines of its surrounding countries. For centuries the Polish kitchen has been the arena for competing with France and Italy. It is rich in meat, especially pork, chicken and beef, winter vegetables (cabbage in the dish bigos) and spices. Pasta is also featured in many Polish dishes, most notably are pierogi. Polish national cuisine shares some similarities with other European traditions. Generally speaking, Polish cuisine is hearty. The preparation of traditional cuisine generally is time intensive and Poles allow themselves a generous amount of time to prepare and enjoy their festive meals, with some meals (like Christmas Eve or Easter breakfast) taking a number of days to prepare in their entirety. It is worth noting that most regions of Poland have their own local gastronomic traditions and distinctive flavours.[141]

Notable foods in Polish cuisine include: soups – rosół, barszcz, żurek, krupnik, kapuśniak, zupa pomidorowa, zupa ogórkowa, zupa grzybowa, flaczki (tripe soup); pierogi, kiełbasa, gołąbki, oscypek, kotlet schabowy, kotlet mielony, bigos, various potato dishes, kanapka, zapiekanka, and many more. Traditional Polish desserts include pączki, faworki, gingerbread, babka and others.

Characteristic dishes are soured milk, buttermilk, kefir, gherkin, pickled cucumber, sauerkraut.

Sports

Football (soccer) is the country's most popular sport, with a rich history of international competition.[142] [143] Track and field, basketball, boxing, ski jumping, fencing, handball, ice hockey, swimming, volleyball, and weightlifting are other popular sports. The golden era of football in Poland occurred throughout the 1970s and went on until the early 1980s when the Polish national football team achieved their best results in any FIFA World Cup competitions finishing 3rd place in the 1974 and 1982 editions. The team won a gold medal in football at the 1972 Summer Olympics and also won two silver medals in 1976 and 1992. Poland, along with Ukraine, hosted the UEFA European Football Championship in 2012.[144]

The Polish men's national volleyball team is ranked 4th in the world and the women's volleyball team is ranked 15th. Mariusz Pudzianowski is a highly successful strongman competitor and has won more World's Strongest Man titles than any other competitor in the world, winning the event in 2008 for the fifth time. The first Polish Formula One driver, Robert Kubica, has brought awareness of Formula One Racing to Poland. Poland has made a distinctive mark in motorcycle speedway racing thanks to Tomasz Gollob, a highly successful Polish rider. The national speedway team of Poland is one of the major teams in international speedway and is very successful in various competitions.[145]

The Polish mountains are an ideal venue for hiking, skiing and mountain biking and attract millions of tourists every year from all over the world.[88] Baltic beaches and resorts are popular locations for fishing, canoeing, kayaking and a broad-range of other water-themed sports.

International rankings

The following are links to international rankings of Poland.

Index	Rank	Countries reviewed
Human Development Index 2011	39th	187
Corruption Perceptions Index 2012	41st	176
OECD Working time	2nd	27
Index of Economic Freedom 2012	64th	179
Privacy International Yearly Privacy ranking of countries, 2007	19th	45
Reporters Without Borders Press Freedom Index 2011–2012	24th	179
UNICEF	14th	21
Networked Readiness Index 2010–2011	62nd	138
OICA Automobile Production 2010	18th	48

See also

- Outline of Poland
- Polish migration to the United Kingdom

Notes

a Numerous sources state that Polish Army was the Allies' fourth biggest fighting contingent. Steven J. Zaloga and Richard Hook write that "by the war's end the Polish Army was the fourth largest contingent of the Allied coalition after the armed forces of the Soviet Union, the United States and the United Kingdom."[146] Jerzy Jan Lerski writes "All in all, the Polish units, although divided and controlled by different political orientation, constituted the fourth largest Allied force, after the America, British and Soviet Armies."[147] M. K. Dziewanowski has noted that "if

Polish forces fighting in the east and west were added to the resistance fighters, Poland had the fourth largest Allied army in the war (after the USSR, the U.S. and Britain)".[148]

The claim of the fourth biggest Ally needs to be reconsidered, however. Throughout the war, Poland's position varied from the 2nd biggest Ally (after the fall of France, when Polish army outnumbered the French) to perhaps the 5th at the end of it (after the USA, Soviet Union, China and Britain). Please, see the analysis in Polish contribution to World War II.

b Sources vary with regards to what was the largest resistance movement during World War II. The confusion often stems from the fact that as war progressed, some resistance movements grew larger – and other diminished. Polish territories were mostly freed from Nazi German control in the years 1944–1945, eliminating the need for their respective (anti-Nazi) partisan forces (in Poland (although the cursed soldiers continued to fight against the Soviets). Several sources note that Polish Armia Krajowa was the largest resistance movement in Nazi-occupied Europe. Norman Davies wrote: "Armia Krajowa (Home Army), the AK, which could fairly claim to be the largest of European resistance";[149] Gregor Dallas wrote "Home Army (Armia Krajowa or AK) in late 1943 numbered around 400000, making it the largest resistance organization in Europe";[150] Mark Wyman wrote "Armia Krajowa was considered the largest underground resistance unit in wartime Europe".[151] Certainly, Polish resistance was the largest resistance till German invasion of Yugoslavia and invasion of the Soviet Union in 1941. After that point, the numbers of Soviet partisans and Yugoslav partisans begun growing rapidly. The numbers of Soviet partisans quickly caught up and were very similar to that of the Polish resistance.[152] [153] The numbers of Tito's Yugoslav partisans were roughly similar to those of the Polish and Soviet partisans in the first years of the war (1941–1942), but grew rapidly in the latter years, outnumbering the Polish and Soviet partisans by 2:1 or more (estimates give Yugoslavian forces about 800,000 in 1945, to Polish and Soviet forces of 400,000 in 1944).[153] [154]

References

[1] Central Statistical Office – Page 18. "Wyniki Narodowego Spisu Powszechnego Ludności i Mieszkań 2011 Podstawowe informacje o sytuacji demograficzno-społecznej ludności Polski oraz zasobach mieszkaniowych" (http://www.stat.gov.pl/cps/rde/xbcr/gus/PUBL_lu_nps2011_wyniki_nsp2011_22032012.pdf). .

[2] "GUS – Population as of 30.06.2010" (http://www.stat.gov.pl/gus/5840_655_ENG_HTML.htm). Stat.gov.pl. . Retrieved 2011-05-26.

[3] "Poland" (http://www.imf.org/external/pubs/ft/weo/2012/01/weodata/weorept.aspx?pr.x=25&pr.y=3&sy=2009&ey=2012&scsm=1&ssd=1&sort=country&ds=.&br=1&c=964&s=NGDPD,NGDPDPC,PPPGDP,PPPPC,LP&grp=0&a=). International Monetary Fund. . Retrieved 2012-04-20.

[4] "Table 1 – Human Development Index and its components" (http://hdr.undp.org/en/media/HDR_2010_EN_Table1.pdf). *Human Development Index 2010*. Human Development Reports. . Retrieved 4 November 2010.

[5] "Concise Statistical Yearbook of Poland, 2008" (http://www.stat.gov.pl/cps/rde/xbcr/gus/PUBL_maly_rocznik_statystyczny_2008.pdf) (PDF). Central Statistical Office (Poland). 28 July 2008. . Retrieved 2008-08-12.

[6] NationMaster.com 2003–2007, Poland, Facts and figures (http://www.nationmaster.com/country/pl-poland)

[7] Lukowski, Jerzy; Zawaszki, Hubert (2001). *A Concise History of Poland* (First Edition ed.). University of Stirling Libraries – Popular Loan (Q 43.8 LUK): Cambridge University Press. p. 3. ISBN 0-521-55917-0.

[8] "Poland – UNESCO World Heritage Centre" (http://whc.unesco.org/en/statesparties/pl). Whc.unesco.org. . Retrieved 2012-02-06.

[9] (English) "Human Development Index and its components" (http://hdr.undp.org/en/media/HDR_2010_EN_Table1.pdf). *hdr.undp.org*. . Retrieved 2011-08-27.

[10] **Polenia** by Thietmar of Merseburg *Chronicle*, 1002. (German: *Polen*)

[11] **Polani** by John Canaparius, *Vita sancti Adalberti episcopi Pragensis*, or *Life of St. Adalbert of Prague*, 999.

[12] "fr. pal, pele, altd. pal, pael, dn. pael, sw. pale, isl. pall, bre. pal, peul, it. polo, pole, pila, [in:] A dictionary of the Anglo-Saxon languages. Joseph Bosworth. S.275.; planus, plain, flat; from Indo- Germanic pele, flat, to spread, also the root of words like plan, floor, and field. [in:] John Hejduk. Soundings. 1993. p. 399"; "the root pele is the source of the English words "field" and "floor". The root "plak" is the source of the English word "flake" [in:] Loren Edward Meierding. Ace the Verbal on the SAT. 2005. p. 82

[13] Teeple, J. B. (2002). *Timelines of World History*. Publisher: DK Adult.

[14] Jerzy Wyrozumski – *Historia Polski do roku 1505* (History of Poland until 1505), Państwowe Wydawnictwo Naukowe (Polish Scientific Publishers PWN), Warszawa 1986, ISBN 978-83-01-03732-1

[15] (English) Norman Davies (1996). *Europe: a history*. Oxford University Press. p. 428. ISBN 0-19-820171-0. "By 1490 the Jagiellons controlled Poland-Lithuania, Bohemia, and Hungary, but not the Empire."

[16] " Jagiellon dynasty (European history) (http://www.britannica.com/EBchecked/topic/299331/Jagiellon-dynasty)". Encyclopædia Britannica.

[17] Davies (2007). *Warfare, State and Society on the Black Sea Steppe,1500–1700.* (http://books.google.com/books?id=XH4hghHo1qoC&pg=PA17&dq&hl=en#v=onepage&q=&f=false). p.17.

[18] " The Crimean Tatars and their Russian-Captive Slaves (http://www.econ.hit-u.ac.jp/~areastd/mediterranean/mw/pdf/18/10.pdf)" (PDF). Eizo Matsuki, *Mediterranean Studies Group at Hitotsubashi University.*

[19] Józef Andrzej Gierowski – *Historia Polski 1505–1764* (History of Poland 1505–1764), Państwowe Wydawnictwo Naukowe (Polish Scientific Publishers PWN), Warszawa 1986, ISBN 978-83-01-03732-1

[20] Józef Andrzej Gierowski – *Historia Polski 1505–1764* (History of Poland 1505–1764), p. 105-173

[21] " Poland – The 17th-century crisis (http://www.britannica.com/EBchecked/topic/466681/Poland/28190/The-17th-century-crisis)". Britannica Online Encyclopedia.

[22] Józef Andrzej Gierowski – *Historia Polski 1505–1764* (History of Poland 1505–1764), p. 174-301

[23] Józef Andrzej Gierowski – *Historia Polski 1764–1864* (History of Poland 1764–1864), Państwowe Wydawnictwo Naukowe (Polish Scientific Publishers PWN), Warszawa 1986, ISBN 978-83-01-03732-1, p. 1-74

[24] Józef Andrzej Gierowski – *Historia Polski 1764–1864* (History of Poland 1764–1864), p. 74-101

[25] " Poland – Independence Won and Lost, 1914–45. (http://countrystudies.us/poland/13.htm)". Library of Congress Country Studies.

[26] Frątczak, Sławomir Z. (2005). "Cud nad Wisłą" (http://web.archive.org/web/20070708173639/http://www.glos.com.pl/Archiwum_nowe/Rok+2005/032/strona/Cud.html) (in Polish). *Głos* (32/2005). Archived from the original (http://www.glos.com.pl/Archiwum_nowe/Rok+2005/032/strona/Cud.html) on 2007-07-08. . Retrieved June 18, 2006.

[27] Bitter glory: Poland and its fate, 1918 to 1939; p.179

[28] " Russian parliament condemns Stalin for Katyn massacre (http://www.bbc.co.uk/news/world-europe-11845315)". BBC News. November 26, 2010

[29] At the siege of Tobruk

[30] including the capture of the monastery hill at the Battle of Monte Cassino

[31] **(English)** Lynne Olson & Stanley Cloud. 2003. A Question of Honor. The Kosciuszko Squadron: Forgotten Heroes of World War II. New York: Knopf.

[32] Peszke, Michael Alfred (February 1999). *Poland's Navy, 1918–1945.* Hippocrene Books. p. 37. ISBN 978-0-7818-0672-5.

[33] Stanisław Salmonowicz, *Polskie Państwo Podziemne*, Wydawnictwa Szkolne i Pedagogiczne, Warszawa, 1994, ISBN 978-83-02-05500-3, p.37

[34] The Warsaw Rising (http://www.polandinexile.com/rising.htm), polandinexile.com

[35] Richard J. Kozicki, Piotr Wróbel (eds), *Historical Dictionary of Poland, 966–1945*, Greenwood Press, 1996, ISBN 978-0-313-26007-0, Google Print, p.34 (http://books.google.com/books?id=S6aUBuWPqywC&pg=PA34&ots=VwP7o88St6&dq=Zygmunt+Berling+Warsaw+Uprising&sig=CuHYk98fb5OPXX7ZTyEiFKnuc5I)

[36] Yad Vashem, The Holocaust Martyrs' and Heroes' Remembrance Authority, Righteous Among the Nations – per Country & Ethnic Origin January 1, 2009. Statistics (http://www1.yadvashem.org/righteous_new/statistics.html)

[37] Richard C. Lukas, *Out of the Inferno: Poles Remember the Holocaust*, University Press of Kentucky 1989 – 201 pages. Page 13; also in Richard C. Lukas, *The Forgotten Holocaust: The Poles Under German Occupation, 1939–1944*, University Press of Kentucky, 1986, Google Print, p.13 (http://books.google.ca/books?id=lz9obsxmuW4C&pg=PA13&dq="&sig=ACfU3U0SGgyvqSbL4bypepYoO_CbYc_N_w).

[38] " European Refugee Movements After World War Two (http://www.bbc.co.uk/history/worldwars/wwtwo/refugees_01.shtml)". BBC – History.

[39] **(English)** "Merkel honours Polish freedom struggle and Tiananmen victims" (http://www.thelocal.de/politics/20090604-19717.html). *www.thelocal.de*. AFP. . Retrieved 2009-09-28.

[40] Arthur Bliss Lane *I saw Poland betrayed: An American Ambassador Reports to the American People.* Indianapolis: The Bobbs-Merrill Company, 1948.

[41] **(Polish)** Polska. Historia (http://encyklopedia.pwn.pl/haslo.php?id=4575043) PWN Encyklopedia. Retrieved 11 July 2005.

[42] "Real GDP growth in CEECs" (http://transitioneconomies.blogspot.com/2006/05/real-gdp-growth-in-ceecs.html). Transitioneconomies.blogspot.com. 2006-05-28. . Retrieved 2009-05-06.

[43] "WHY POLAND?" (http://polisci.lsa.umich.edu/documents/jjackson/chapt1.pdf) (PDF). . Retrieved 2009-07-08.

[44] "Europe's border-free zone expands" (http://news.bbc.co.uk/1/hi/7153490.stm). BBC News. 21 December 2007. . Retrieved 28 July 2011.

[45] Dyrekcja Generalna Lasów Państwowych, Raport o stanie lasów w Polsce (Report on the Status of Forests in Poland), (http://bip.lasy.gov.pl/pl/bip/px_~raport_o_stanie_lasow_2011.pdf?page_opener=http://bip.lasy.gov.pl/pl/bip/raporty_i_prognozy) *Centrum Informacyjne Lasów Państwowych.* **(Polish)**

[46] "Poland.pl – White Stork – About White Stork" (http://storks.poland.pl/about_stork/index.htm). Storks.poland.pl. . Retrieved 2009-05-06.

[47] **(Polish)** "Wybrzeże Morza Bałtyckiego" (http://zalewszczecinski.net/akweny/wybrzeze-morza-baltyckiego). *www.zalewszczecinski.net.* . Retrieved 2009-11-16.

[48] "News Poland | Minorities make the news" (http://www.newpolandexpress.pl/polish_news_story-3634-minorities_make_the_news_.php). Newpolandexpress.pl. 2011-10-15. . Retrieved 2011-11-02.

[49] "Transgender woman poised for seat in Poland's new parliament" (http://www.telegraph.co.uk/news/worldnews/europe/poland/8817279/Transgender-woman-poised-for-seat-in-Polands-new-parliament.html). London: Telegraph. 2011-10-10. . Retrieved 2011-11-02.

[50] Davies, Norman (1996). *Europe: A History* (http://books.google.com/books?id=jrVW9W9eiYMC&pg=PA699). Oxford University Press. p. 699. ISBN 0-19-820171-0. .

[51] Grushenko, Kateryna (12 November 2010). "Polish representative: 'Poland is ready to help Ukraine as long as you are interested'" (http://www.kyivpost.com/news/guide/world-in-uktaine/detail/89674/). Kyiv Post. . Retrieved 28 July 2011.

[52] "Bordering on madness: Belarus mistreats its Polish minority" (http://economist.com/displaystory.cfm?story_id=4085710). The Economist. 16 June 2005. . Retrieved 28 July 2011.

[53] "Jerzy Buzek elected President of the European Parliament" (http://www.europarl.europa.eu/sides/getDoc.do?type=IM-PRESS&reference=20090713IPR58063&language=EN). European Parlament. 2009-07-14. . Retrieved 2010-12-12.

[54] (Polish) "Strategia Bezpieczeństwa Narodowego RP" (http://www.wp.mil.pl/pliki/File/zalaczniki_do_stron/SBN_RP.pdf) (PDF). *www.wp.mil.pl*. . Retrieved 2008-09-26.

[55] Day, Matthew (5 August 2008). "Poland ends army conscription" (http://www.telegraph.co.uk/news/worldnews/europe/poland/2505447/Poland-ends-army-conscription.html). London: Telegraph. . Retrieved 28 July 2011.

[56] "Polska zakończyła udział w misjach po auspicjami ONZ – Wiadomości z kraju i ze świata – Gazeta Prawna – Partner pracodawcy, narzędzie specjalisty" (http://www.gazetaprawna.pl/wiadomosci/artykuly/382706,polska_zakonczyla_udzial_w_misjach_po_auspicjami_onz.html). Gazetaprawna.pl. 2009-12-31. . Retrieved 2011-11-02.

[57] "Accident Database" (http://www.airdisaster.com/cgi-bin/view_details.cgi?date=04102010&reg=101&airline=Polish+Air+Force). AirDisaster.com. . Retrieved 2010-12-12.

[58] "Senior Polish figures killed in plane crash" (http://news.bbc.co.uk/2/hi/europe/8613395.stm). BBC. 2010-04-11. .

[59] "– 15 tys. zimowych mundurów trafi do jednostek" (http://www.policja.pl/portal/pol/1/59960/15_tys_zimowych_mundurow_trafi_do_jednostek.html). Policja.pl. 30 September 2009. . Retrieved 28 July 2011.

[60] "Nowe radiowozy dla policji" (http://moto.onet.pl/1545847,1,nowe-radiowozy-dla-policji,artykul.html?node=2). Moto.onet.pl. 3 March 2009. . Retrieved 28 July 2011.

[61] "Country and Lending Groups | Data" (http://data.worldbank.org/about/country-classifications/country-and-lending-groups#High_income). Data.worldbank.org. . Retrieved 2010-11-09.

[62] "How Poland became only EU nation to avoid recession" (http://articles.cnn.com/2010-06-29/world/poland.economy.recession_1_poland-transition-economies-eastern-europe?_s=PM:WORLD). CNN. 29 June 2010. . Retrieved 28 July 2011.

[63] "Central Europe Risks Downgrades on Worsening Finances (Update1)" (http://www.bloomberg.com/apps/news?pid=newsarchive&sid=apz_BEuHrNpI). Bloomberg.com. 2009-09-21. . Retrieved 2010-01-27.

[64] "Zloty to Gain, Says LBBW, Most Accurate Forecaster (Update1)" (http://www.bloomberg.com/apps/news?pid=newsarchive&sid=a7k1g_PkuROs#). Bloomberg.com. 2009-10-09. . Retrieved 2010-01-27.

[65] "Banking Sector in Poland | Thomas White International" (http://www.thomaswhite.com/explore-the-world/emerging-market-spotlight/poland-banking.aspx). Thomaswhite.com. 2011-09-16. .

[66] " Poland in the Lead (http://www.warsawvoice.pl/archiwum.phtml/5473/)", *The Warsaw Voice*, September 2002. Retrieved on August 11, 2007.

[67] "Polish Information and Foreign Investment Agency. News" (http://www.paiz.gov.pl/nowosci/?id_news=1297&lang_id=12). *www.paiz.gov.pl*. . Retrieved 2008-09-26.

[68] "GDP per capita in PPS" (http://epp.eurostat.ec.europa.eu/tgm/table.do?tab=table&init=1&plugin=1&language=en&pcode=tsieb010). Eurostat. . Retrieved 2010-06-25.

[69] Bloomberg Businessweek. 2 December 2011. Official: Poland to be ready for euro in 4 years (http://www.businessweek.com/ap/financialnews/D9RCGVPG2.htm)

[70] Schwab, Klaus. "The Global Competitiveness Report 2010-2011" (http://www3.weforum.org/docs/WEF_GlobalCompetitivenessReport_2010-11.pdf). World Economic Forum. . Retrieved 25 April 2011.

[71] Shapiro, Robert J.. "Foreign Direct Investments in Developing Nations: Issues in Telecommunications and the Modernization of Poland" (http://www.investmentsecurity.org/wp-content/uploads/2011/02/CEIS-FDI-Report_Apr-2011.pdf). CEIS. . Retrieved 27 April 2011.

[72] "Waking up to the new economy: Ernst & Young's 2010 European attractiveness survey" (http://www.ey.com/Publication/vwLUAssets/Attractiveness_survey_2010_EU/$FILE/Attractiveness_survey_2010_EU.pdf). Ernst & Young. . Retrieved 25 April 2011.

[73] "Communication on the average monthly salary in enterprise sector excluding payments from profit awards in December 2010" (http://finanse.wp.pl/kat,58434,title,GUS-podal-dane-o-wynagrodzeniach,wid,13051878,wiadomosc.html?ticaid=1ba11) (in (Polish)). Stat.gov.pl. . Retrieved 2010-11-04.

[74] "OECD Economic Outlook No. 82 – Poland" (http://web.archive.org/web/20070328163217/http://www.oecd.org/dataoecd/6/32/20213254.pdf) (PDF). Archived from the original (http://www.oecd.org/dataoecd/6/32/20213254.pdf) on 2007-03-28. . Retrieved 2010-04-12.

[75] "Statistic Office of Poland(GUS)" (http://www.stat.gov.pl/bdr_n/app/wybrane_cechy.wymiary) (in (Polish)). Stat.gov.pl. . Retrieved 2010-11-19.

[76] "Poles getting rich quick..." (http://www.thenews.pl/business/artykul141422_poles-getting-rich-quick-.html). Polskie Radio, thenews.pl. . Retrieved 2010-10-13.

[77] Jędrzej Bielecki. "Polacy są w światowej czołówce bogacących się narodów" (http://forsal.pl/artykuly/458053,polacy_sa_w_swiatowej_czolowce_bogacacych_sie_narodow.html). Dziennik Gazeta Prawna. . Retrieved 2010-10-13.

[78] http://www.cespi.it/WPMIG/Country%20mig-POLAND.pdf

[79] "Topic Galleries" (http://www.chicagotribune.com/news/sns-rt-us-poland-immigrantstre7bs11w-20111229,0,2982543.story). chicagotribune.com. . Retrieved 2012-02-06.

[80] Kasat Sp. z o.o. (2008-12-31). "Imigranci w Polsce 2008 – eGospodarka.pl – Raporty i prognozy" (http://www.egospodarka.pl/36792,Imigranci-w-Polsce-2008,1,39,1.html). eGospodarka.pl. . Retrieved 2012-02-06.

[81] "Coraz więcej Azjatów emigruje do Polski" (http://fakty.interia.pl/prasa/news/coraz-wiecej-azjatow-emigruje-do-polski,1326394,16). Interia.pl. . Retrieved 2010-10-13.

[82] "Eurostat February 2008 – Euro area unemployment stable at 7.1%" (http://web.archive.org/web/20080726044250/http://epp.eurostat.ec.europa.eu/pls/portal/docs/PAGE/PGP_PRD_CAT_PREREL/PGE_CAT_PREREL_YEAR_2008/PGE_CAT_PREREL_YEAR_2008_MONTH_04/3-01042008-EN-AP.PDF). Web.archive.org. 26 July 2008. Archived from the original (http://epp.eurostat.ec.europa.eu/pls/portal/docs/PAGE/PGP_PRD_CAT_PREREL/PGE_CAT_PREREL_YEAR_2008/PGE_CAT_PREREL_YEAR_2008_MONTH_04/3-01042008-EN-AP.PDF) on 2008-07-26. . Retrieved 28 July 2011.

[83] "Polish economy seen as stable and competitive" (http://www.wbj.pl/article-51029-polish-economy-seen-as-stable-and-competitive.html). Warsaw Business Journal. 9 September 2010. . Retrieved 28 July 2011.

[84] ATLAS AMONGST THE BEST POLISH BRANDS (http://dziennik.dziennikbudowy.pl/ang.nsf/0/2CE31E39A26A64F3C1256C31002C1FA4?OpenDocument). Retrieved 21 November 2010.

[85] "fDi: Poland Primed for Golden Decade" (http://www.ginannebrownell.com/2010/10/fdi-poland-primed-for-golden-decade/). GinanneBrownell.com. 8 October 2010. . Retrieved 28 July 2011.

[86] (Polish) "Lista 500 największych polskich firm" (http://www.lista500.polityka.pl/rankings/show). *www.lista500.polityka.pl.* . Retrieved 2011-08-27.

[87] (English) "Travel And Tourism in Poland" (http://www.euromonitor.com/Travel_And_Tourism_in_Poland). *www.euromonitor.com.* . Retrieved 2009-10-12.

[88] (English) "UNTWO World Tourism Barometer, Vol.5 No.2" (http://www.tourismroi.com/Content_Attachments/27670/File_633513750035785076.pdf). *www.tourismroi.com.* . Retrieved 2009-10-12.

[89] "Use of renewable fuel in Central and Eastern Europe" (http://www.frost.com/prod/servlet/cpo/205103205.htm). Frost.com. 1 March 2009. . Retrieved 28 July 2011.

[90] "EU Commission – Energy factsheet P74" (http://ec.europa.eu/energy/publications/statistics/doc/2010_energy_transport_figures.pdf) (PDF). . Retrieved 2011-07-28.

[91] "National Road Rebuilding Program (Polish)" (http://bip.mswia.gov.pl/portal/bip/175/17721/Narodowy_Program_Przebudowy_Drog.html). Bip.mswia.gov.pl. 2006-02-16. . Retrieved 2011-07-28.

[92] "Super pociągi zamiast autostrad (Polish)" (http://www.tvn24.pl/12690,1634983,0,1,super-pociagi-zamiast-autostrad,wiadomosc.html). *TVN24*. December 23, 2009. . Retrieved December 25, 2009.

[93] (English) Richard Francis Mould (1993). *A century of X-rays and radioactivity in medicine: with emphasis on photographic records of the early years* (http://books.google.com/?id=IXPz7bVR7g0C&printsec=frontcover&dq=A+century+of+x-rays+and+radioactivity+in+medicine:&q=). p. 19. ISBN 978-0-7503-0224-1. .

[94] Newswire Poland Emerges as the European R&D Hub Despite Favorable Conditions in Asia Pacific (http://www.newswiretoday.com/news/26490/)

[95] Polish Information and Foreign Investment Agency Poland – R&D centre (http://web.archive.org/web/20080219145055/http://www.paiz.gov.pl/index/?id=7b7a53e239400a13bd6be6c91c4f6c4e)

[96] (English) KPMG Sp. z o.o.. "Why Poland?" (http://www.paiz.gov.pl/files/?id_plik=7513). *www.paiz.gov.pl*. p. 3. . Retrieved 2011-08-27. "Over 80% of foreign investors see the results of their investments to date as positive or very positive and none of the studied companies reported a negative opinion."

[97] (English) "Key data on IT and telecoms market in Poland, 2004–2006" (http://web.archive.org/web/20061108212651/http://www.sat.org.au/reviews/articles_pl_middle_ages.htm). *www.itandtelecompoland.com*. Archived from the original (http://www.itandtelecompoland.com/index.php?item=2) on 2006-11-08. . Retrieved 2008-09-24.

[98] *The Expulsion of 'German' Communities from Eastern Europe at the end of the Second World War* (http://cadmus.iue.it/dspace/bitstream/1814/2599/1/HEC04-01.pdf), Steffen Prauser and Arfon Rees, European University Institute, Florense. HEC No. 2004/1. p.29

[99] (English) Michał Buchowski, Katarzyna Chlewińska. "Tolerance and Cultural Diversity Discourses in Poland" (http://www.eui.eu/Projects/ACCEPT/Documents/Research/wp1/ACCEPTPLURALISMWp1BackgroundreportPoland.pdf). *www.eui.eu.* . Retrieved 2011-08-27.

[100] (English) Jan Repa (2007-01-05). "Poles return to Russian language" (http://news.bbc.co.uk/2/hi/europe/6233821.stm). *news.bbc.co.uk.* . Retrieved 2011-08-27. "In former satellite countries like Hungary or Poland, knowledge of Russian dwindled rapidly – to be replaced by English and German."

[101] Doughty, Steve (2006-04-25). "UK lets in more Poles than there are in Warsaw" (http://www.dailymail.co.uk/news/article-384121/UK-lets-Poles-Warsaw.html). London: Dailymail.co.uk. . Retrieved 2010-04-12.

[102] (English) Alexi Mostrous, Christine Seib (February 16, 2008). "Tide turns as Poles end great migration" (http://www.timesonline.co.uk/tol/news/uk/article3378877.ece). *www.timesonline.co.uk* (London). . Retrieved 2011-08-27. "The Times has established that, for the first

time since they began arriving en masse four years ago, more UK-based Poles are returning to their homeland than are entering Britain."

[103] "Polish Diaspora (Polonia) Worldwide" (http://culture.polishsite.us/articles/art79fr.htm). Culture.polishsite.us. . Retrieved 2010-04-12.

[104] "Centers of Polish Immigration in the World – USA and Germany" (http://culture.polishsite.us/articles/art90fr.htm). Culture.polishsite.us. 2003-03-15. . Retrieved 2010-04-12.

[105] "Maly Rocznik Statystyczny Polski 2009" (http://www.stat.gov.pl/cps/rde/xbcr/gus/PUBL_oz_maly_rocznik_statystyczny_2009.pdf) (in (Polish)) (PDF). . Retrieved 2009-09-26.

[106] "*94% Polaków wierzy w Boga*" (http://www.ekumenizm.pl/content/article/20080925183042429.htm). Ekumenizm.pl. 2008-09-25. . Retrieved 2010-04-12.

[107] Weekly Mass Attendance of Catholics in Nations with Large Catholic Populations, 1980–2008 (http://web.archive.org/web/20080214110918/http://cara.georgetown.edu/bulletin/international.htm) – World Values Survey (WVS)

[108] (Polish) Centrum Badania Opinii Społecznej (*Centre for Public Opinion Research (Poland)* CBOS). Komunikat z badań; Warszawa, Marzec 2005. Co łączy Polaków z parafią? (http://www.cbos.pl/SPISKOM.POL/2005/K_049_05.PDF) Preface. Retrieved 2007-12-14.

[109] Wilde, Robert. "Pope John Paul II 1920–2005" (http://europeanhistory.about.com/od/religionandthought/a/biojohnpaulii.htm). About.com. . Retrieved 2009-01-01.

[110] Domínguez, Juan: 2005

[111] "Pope John Paul II and Communism" (http://www.religion-cults.com/pope/communism.htm). Public domain text. May be distributed freely. No rights reserved.. . Retrieved 2009-01-01.

[112] (Polish) Dr Zbigniew Pasek, Jagiellonian University, "Wyznania religijne" (http://web.archive.org/web/20061128165649/http://www.religioznawstwo.uj.edu.pl/syllabusy/pasek-wrwp.rtf). Archived from the original (http://www.religioznawstwo.uj.edu.pl/syllabusy/pasek-wrwp.rtf) on November 28, 2006. . Retrieved 2007-09-15. Further reading: Ustawa o gwarancjach wolności sumienia i wyznania z dnia 17 V 1989 z najnowszymi nowelizacjami z 1997 roku.

[113] (Polish) Michał Tymiński, "Kościół Zielonoświątkowy" (http://web.archive.org/web/20050102151031/http://www.kz.pl/index.php?p=13&id=3&i=8). Archived from the original (http://www.kz.pl/index.php?p=13&id=3&i=8) on January 2, 2005. . Retrieved 2007-09-14.

[114] (Polish) Dr. Paweł Borecki, "Opinia prawna dotycząca religii w szkole" (http://www.racjonalista.pl/kk.php/s,5534). Kateda Prawa Wyznaniowego Uniwersytetu Warszawskiego. . Retrieved 2007-09-14.

[115] (Polish) Wirtualna Polska, Wiadomości. "Polacy przeciwni wliczaniu ocen z religii do średniej" (http://wiadomosci.wp.pl/kat,9911,wid,9125933,wiadomosc.html?ticaid=1478d). . Retrieved 2007-09-14.

[116] (Polish) Olga Szpunar, "Dorośli chcą religii w szkole" (http://miasta.gazeta.pl/krakow/1,35798,4360977.html). Gazeta Wyborcza Kraków. . Retrieved 2007-09-15.

[117] "Poland Guide: The Polish health care system, An introduction: Poland's health care is based on a general" (http://www.justlanded.com/english/Poland/Poland-Guide/Health/The-Polish-health-care-system). Justlanded.com. . Retrieved 2011-07-28.

[118] "Polish hospitals" (http://polandpoland.com/polish_hospitals.html). Polandpoland.com. . Retrieved 2011-07-28.

[119] "The Quality of Medical Treatment and Surgery in Poland" (http://www.articlesbase.com/medicine-articles/the-quality-of-medical-treatment-and-surgery-in-poland-602240.html). Articlesbase.com. . Retrieved 2011-07-28.

[120] "WHO | Poland" (http://www.who.int/countries/pol/en/). Who.int. 2011-05-17. . Retrieved 2011-07-28.

[121] OECD (2009). "The impact of the 1999 education reform in Poland" (http://www.pisa.oecd.org/dataoecd/50/26/45721631.doc). . Retrieved 2010-09-17.

[122] (English) "Range of rank on PISA 2006 science scale" (http://www.oecd.org/dataoecd/42/8/39700724.pdf) (PDF). *www.oecd.org*. . Retrieved 2008-09-24.

[123] Adam Zamoyski, The Polish Way: A Thousand Year History of the Poles and Their Culture (http://www.amazon.co.uk/Polish-Way-Thousand-History-Culture/dp/978-0-7818-0200-0). Published 1993, Hippocrene Books, Poland, ISBN 978-0-7818-0200-0

[124] Ministry of Foreign Affairs of Poland, 2002–2007, AN OVERVIEW OF POLISH CULTURE. (http://www.poland.gov.pl/Culture,484.html) Access date 12-13-2007.

[125] (English) "Nicolaus Copernicus" (http://www.britannica.com/EBchecked/topic/136591/Nicolaus-Copernicus). *www.britannica.com*. . Retrieved 2008-10-10.

[126] (French) Rey Alain (1993). *Le petit Robert 2 : (dictionnaire universel des noms propres, alphabétique et analogique)*. INIST-CNRS, Cote INIST : L 22712: Le Robert, Paris, FRANCE. ISBN 978-2-85036-210-1.

[127] (English) Michael Kennedy, ed. (2004). *The Concise Oxford dictionary of music*. Oxford University Press. ISBN 978-0-19-860884-4. p. 141

[128] (French) "Maria Sklodowska. La jeunesse" (http://mariecurie.science.gouv.fr/portrait/portrait1_1.php). *mariecurie.science.gouv.fr*. . Retrieved 2008-10-10.

[129] (Polish) Koca, B. (2006). "Polish Literature – The Middle Ages (Religious writings)" (http://web.archive.org/web/20061108212651/http://www.sat.org.au/reviews/articles_pl_middle_ages.htm). Archived from the original (http://www.sat.org.au/reviews/articles_pl_middle_ages.htm) on 2006-11-08. . Retrieved 10 December 2006.

[130] (English) Zdzislaw Najder (1998). "Profiles – Joseph Conrad" (http://www.culture.pl/en/culture/artykuly/os_conrad_joseph). *www.culture.pl*. . Retrieved 2008-09-30.

[131] http://www.coe.int/t/dg3/children/corporalpunishment/pdf/EnglishQuestionAnswer_en.pdf

[132] "Dr. Sławomir Łodziński, Helsinki Foundation for Human Rights, "The Protection of National Minorities in Poland"" (http://www.minelres.lv/reports/poland/poland_NGO.htm). Minelres.lv. . Retrieved 2011-07-28.

[133] http://www.visionofhumanity.org/wp-content/uploads/2011/10/2011GPIMethodologyResultsFindings.pdf

[134] **(English)** "The Music Courts of the Polish Vasas" (http://www.semper.pl/muzyczne_dwory_summary.pdf). *www.semper.pl*. p. 244. . Retrieved 2009-05-13.

[135] Busfield, Steve (2006-02-21). "Guardian wins design award" (http://media.guardian.co.uk/site/story/0,,1714643,00.html). *The Guardian* (London). .

[136] "Facts on the Nobel Prize in Literature" (http://nobelprize.org/nobel_prizes/literature/shortfacts.html). Nobelprize.org. 2009-10-05. . Retrieved 2011-07-28.

[137] Adam Gopnik (June 5, 2007). "Szymborska's 'View': Small Truths Sharply Etched" (http://www.npr.org/templates/story/story.php?storyId=10721773). npr.org. . Retrieved December 12, 2010.

[138] **(English)** "Szydłowiec" (http://web.archive.org/web/20060623020259/http://www.szydlowiec.pl/grafika/index/szydl1.pdf). *www.szydlowiec.pl*. p. 9. Archived from the original (http://www.szydlowiec.pl/grafika/index/szydl1.pdf) on June 23, 2006. . Retrieved 2009-04-23.

[139] Many designs imitated the arcaded courtyard and arched loggias of the Wawel palace. **(English)** Michael J. Mikoś. "RENAISSANCE CULTURAL BACKGROUND" (http://www.staropolska.pl/ang/renaissance/Mikos_renaissance/Cultural_r.html). *www.staropolska.pl*. p. 9. . Retrieved 2009-04-23.

[140] **(English)** John Stanley (March–June 2004). "Literary Activities and Attitudes in the Stanislavian Age in Poland (1764–1795): A Social System?" (http://findarticles.com/p/articles/mi_qa3763/is_200403/ai_n9363971/?tag=content;col1). *findarticles.com*. . Retrieved 2009-04-23.

[141] Encyklopedia kuchni: "Kuchnia polska." (http://www.kuchnia.tv/kuchnia_encyklopedia-kuchni_185_kuchnia-polska.html) Kuchnia.tv, ul. W. Sikorskiego 902-758 Warszawa, 2011.

[142] "FIFA World Cup Statistics-Poland" (http://www.fifa.com/worldfootball/statisticsandrecords/associations/association=pol/worldcup/index.html). FIFA. . Retrieved December 12, 2010.

[143] "FIFA Statistics – Poland" (http://www.fifa.com/worldfootball/statisticsandrecords/associations/association=pol/othertournaments/index.html). . Retrieved December 12, 2010.

[144] "Poland hosts Euro 2012!" (http://www.warsaw-life.com/poland/euro-2012). warsaw-life.com. . Retrieved December 12, 2010.

[145] "Speedway World Cup: Poland win 2010 Speedway World Cup" (http://www.worldspeedway.com/artman/publish/article_13423.shtml). worldspeedway.com. . Retrieved December 18, 2010.

[146] Steven J. Zaloga; Richard Hook (21 January 1982). *The Polish Army 1939–45* (http://books.google.com/books?id=AAdYFeW2fnoC&pg=PA3). Osprey Publishing. pp. 3–. ISBN 978-0-85045-417-8. . Retrieved 6 March 2011.

[147] Jerzy Jan Lerski (1996). *Historical dictionary of Poland, 966–1945* (http://books.google.com/books?id=QTUTqE2difgC&pg=PA18). Greenwood Publishing Group. pp. 18–. ISBN 978-0-313-26007-0. . Retrieved 6 March 2011.

[148] E. Garrison Walters (1988). *The other Europe: Eastern Europe to 1945* (http://books.google.com/books?id=64VpSBd7xUcC&pg=PA276). Syracuse University Press. pp. 276–. ISBN 978-0-8156-2440-0. . Retrieved 6 March 2011.

[149] Norman Davies, *God's Playground: A History of Poland*, Columbia University Press, 2005, ISBN 0-231-12819-3, Google Print p.344 (http://books.google.com/books?id=EBpghdZeIwAC&pg=PA344&dq="Armia+Krajowa"+largest&ei=hTrMR_W-G4mWzASVs9GtCQ&sig=iE7xbtRu3rvEsVZZgCeUsqEqj6s)

[150] Gregor Dallas, *1945: The War That Never Ended*, Yale University Press, 2005, ISBN 0-300-10980-6, Google Print, p.79 (http://books.google.com/books?id=LXdVF6LmTa8C&pg=PA79&dq="Armia+Krajowa"+largest&as_brr=3&ei=RjvMR6KnPJPAzAT-ppWvCQ&sig=Ksba8pTs5pu55YiAqseCLy6Kl5k)

[151] Mark Wyman, *DPs: Europe's Displaced Persons, 1945–1951*, Cornell University Press, 1998, ISBN 0-8014-8542-8, Google Print, p.34 (http://books.google.com/books?id=lHNw7MnsmlYC&pg=PA34&dq="Armia+Krajowa"+largest&lr=&as_brr=3&ei=NzzMR_mOIJGSzQSb7cSwCQ&sig=kv3oN5z3YgAgcT8Vgy4aIFRHknE)

[152] See for example: Leonid D. Grenkevich in The Soviet Partisan Movement, 1941–44: A Critical Historiographical Analysis, p.229 or Walter Laqueur in The Guerilla Reader: A Historical Anthology, (New York, Charles Scribiner, 1990, p.233.

[153] Velimir Vukšić (23 July 2003). *Tito's partisans 1941–45* (http://books.google.com/books?id=SLix5hc4WRgC&pg=PA11). Osprey Publishing. pp. 11–. ISBN 978-1-84176-675-1. . Retrieved 1 March 2011.

[154] Anna M. Cienciala, THE COMING OF THE WAR AND EASTERN EUROPE IN WORLD WAR II. (http://web.ku.edu/~eceurope/hist557/lect16.htm), History 557 Lecture Notes

External links

Movie (on-line)

- Animated history of Poland, (PARP, Expo 2010 Shanghai China) (http://www.youtube.com/watch?v=2DrXgj1NwN8)
- Borders of Poland, AD. 990–2008 (http://www.youtube.com/watch?v=eAVVWlUywO0)
- Poland under German occupation 1939–1945 (http://www.youtube.com/watch?v=nSzejCWXKhA)

General

- Poland.gov.en – Polish national portal (http://en.poland.gov.pl/)
- Polish Information (http://www.paiz.gov.pl/index/?lang_id=1)
- Ministry of Foreign Affairs (http://www.msz.gov.pl/)
- Chief of State and Cabinet Members (https://www.cia.gov/library/publications/world-leaders-1/world-leaders-p/poland.html)
- Poland (https://www.cia.gov/library/publications/the-world-factbook/geos/pl.html) entry at *The World Factbook*
- Poland (http://ucblibraries.colorado.edu/govpubs/for/poland.htm) at *UCB Libraries GovPubs*
- History of Ancient Poland (http://www.theancientweb.com/explore/content.aspx?content_id=24)
- Poland (http://www.dmoz.org/Regional/Europe/Poland/) at the Open Directory Project
- Poland profile (http://www.bbc.co.uk/news/world-europe-17753718) from the BBC News
- Poland (http://www.oecd.org/poland/) OECD
- Poland (http://europa.eu/about-eu/countries/member-countries/poland/index_en.htm) EU
- Poland (http://www.britannica.com/EBchecked/topic/466681/Poland) *Encyclopædia Britannica* entry
- Wikimedia Atlas of Poland
- Geographic data related to Poland (http://www.openstreetmap.org/browse/relation/49715) at OpenStreetMap
- Key Development Forecasts for Poland (http://www.ifs.du.edu/ifs/frm_CountryProfile.aspx?Country=PL) from International Futures

Culture

- Commonwealth of Diverse Cultures: Poland's Heritage (http://www.commonwealth.pl/)

Travel

- Poland's Official Travel Website (http://www.poland.travel/en)
- About Poland (http://www.international-travel-tours.com/poland/about-poland.html)
- Photos from Poland (http://www.klub-beskid.com/album-pologne/index.php?lang=english)

Widawa, Łódź Voivodeship

Widawa	
— Village —	

Widawa

Coordinates: 51°26′14″N 18°56′21″E

Country	Poland
Voivodeship	Łódź
County	Łask
Gmina	Widawa
Population	1300

Widawa [vi'dava] is a village in Łask County, Łódź Voivodeship, in central Poland.[1] It is the seat of the gmina (administrative district) called Gmina Widawa. It lies approximately 22 kilometres (14 mi) south-west of Łask and 53 km (33 mi) south-west of the regional capital Łódź.

The village has a population of 1,300.

References

[1] "Central Statistical Office (GUS) - TERYT (National Register of Territorial Land Apportionment Journal)" (http://www.stat.gov.pl/broker/access/prefile/listPreFiles.jspa) (in Polish). 2008-06-01. .

- Jewish history of Widawa (http://www.jewishgen.org/yizkor/pinkas_poland/pol1_00094.html)

Łask

Łask	
 Church in Łask	
 **Coat of arms**	
 Łask	
Coordinates: 51°35′N 19°8′E	
Country	Poland
Voivodeship	Łódź
County	Łask County
Gmina	Gmina Łask
Government	
• **Mayor**	Gabriel Szkudlarek
Area	
• **Total**	15.33 km^2 (**unknown operator: u'strong'** sq mi)
Population (2006)	
• **Total**	18684

• Density	**unknown operator: u'strong'**/km^2 (**unknown operator: u'strong'**/sq mi)
Time zone	CET (UTC+1)
• Summer (DST)	CEST (UTC+2)
Postal code	98-100
Car plates	ELA
Website	http://www.lask.pl

Łask [wask] is a town in central Poland with 18,948 inhabitants (2004). It's the capital of Łask County, and is situated in Łódź Voivodeship (since 1999), previously in Sieradz Voivodeship (1975–1998). 32nd Air Base of the Polish Air Force is located nearby.

History

With the invasion of Poland and the outbreak of the Second World War in September 1939, Łask was occupied by the Wehrmacht and annexed by Nazi Germany. It was administered as part of the county or district (*kreis*) of Lask within Reichsgau Wartheland. Following the arrival of the Red Army and the end of the war, Łask became part of the People's Republic of Poland.

References

- The Lost Jewish Community of Łask [1]

See also

- Lasker (disambiguation)
- Laski (disambiguation)
- Shtetl

References

[1] http://kehilalinks.jewishgen.org/lask/

Łask County

Łask County
Powiat łaski

— County —

Flag

Coat of arms

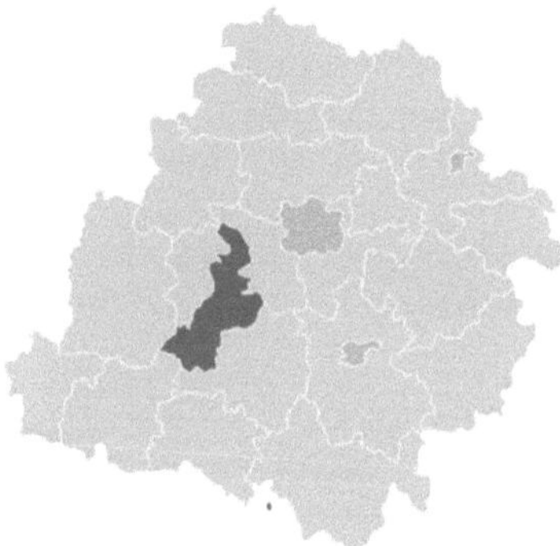

Location within the voivodeship

Division into gminas

Coordinates (Łask): 51°35′N 19°8′E

Country	Poland
Voivodeship	Łódź

Seat	Łask
Gminas	
Area	
• Total	617.38 km^2 (**unknown operator: u'strong'** sq mi)
Population (2006)	
• Total	50874
• Density	**unknown operator: u'strong'**/km^2 (**unknown operator: u'strong'**/sq mi)
• Urban	18684
• Rural	32190
Car plates	ELA
Website	http://www.lask.com.pl

Łask County (Polish: *powiat łaski*) is a unit of territorial administration and local government (powiat) in Łódź Voivodeship, central Poland. It came into being on January 1, 1999, as a result of the Polish local government reforms passed in 1998. Its administrative seat and only town is Łask, which lies 32 kilometres (20 mi) south-west of the regional capital Łódź.

The county covers an area of 617.38 square kilometres (238.4 sq mi). As of 2006 its total population is 50,874, out of which the population of Łask is 18,684 and the rural population is 32,190.

Neighbouring counties

Łask County is bordered by Poddębice County to the north, Pabianice County to the east, Bełchatów County to the south-east, Wieluń County to the south-west, and Sieradz County and Zduńska Wola County to the west.

Administrative division

The county is subdivided into five gminas (one urban-rural and four rural). These are listed in the following table, in descending order of population.

Gmina	Type	Area (km²)	Population (2006)	Seat
Gmina Łask	urban-rural	146.9	28,406	Łask
Gmina Widawa	rural	178.0	7,954	Widawa
Gmina Sędziejowice	rural	120.2	6,523	Sędziejowice
Gmina Buczek	rural	90.8	4,893	Buczek
Gmina Wodzierady	rural	81.4	3,098	Wodzierady

References

- Polish official population figures 2006 [1]

References

[1] http://www.stat.gov.pl/gus/45_655_PLK_HTML.htm

Gmina Widawa

Gmina Widawa Widawa Commune — Gmina —	
 Coat of arms	
Coordinates (Widawa): 51°26′14″N 18°56′21″E	
Country	Poland
Voivodeship	Łódź
County	Łask
Seat	Widawa
Area	
• Total	178.04 km^2 (**unknown operator: u'strong'** sq mi)
Population (2006)	
• Total	7954
• Density	**unknown operator: u'strong'**/km^2 (**unknown operator: u'strong'**/sq mi)
Website	http://www.widawa.finn.pl

Gmina Widawa is a rural gmina (administrative district) in Łask County, Łódź Voivodeship, in central Poland. Its seat is the village of Widawa, which lies approximately 22 kilometres (14 mi) south-west of Łask and 53 km (33 mi) south-west of the regional capital Łódź.

The gmina covers an area of 178.04 square kilometres (68.7 sq mi), and as of 2006 its total population is 7,954.

The gmina contains part of the protected area called Warta-Widawka Landscape Park.

Villages

Gmina Widawa contains the villages and settlements of Brzyków, Chociw, Chrusty, Chrząstawa, Dąbrowa Widawska, Dębina, Górki Grabińskie, Goryń, Grabówie, Izydorów, Józefów Widawski, Kąty, Klęcz, Kocina, Kolonia Zawady, Korzeń, Las Zawadzki, Łazów, Ligota, Lucjanów, Ochle, Ochle-Kolonia, Osieczno, Patoki, Podgórze, Przyborów, Raczynów, Restarzew Cmentarny, Restarzew Środkowy, Rogóźno, Ruda, Sarnów, Sewerynów, Siemiechów, Świerczów, Widawa, Wielka Wieś A, Wielka Wieś B, Wincentów, Witoldów, Wola Kleszczowa, Zabłocie, Zawady and Zborów.

Neighbouring gminas

Gmina Widawa is bordered by the gminas of Burzenin, Konopnica, Rusiec, Sędziejowice, Szczerców, Zapolice and Zelów.

References

- Polish official population figures 2006 [1]
- Jewish history of Widawa [1]

References

[1] http://www.jewishgen.org/yizkor/pinkas_poland/pol1_00094.html

Village

A **village** is a clustered human settlement or community, larger than a hamlet with the population ranging from a few hundred to a few thousand (sometimes tens of thousands). Though often located in rural areas, the term urban village is also applied to certain urban neighbourhoods, such as the East Village in Manhattan, New York City and the Saifi Village in Beirut, Lebanon, as well as Hampstead Village in the London conurbation. Villages are normally permanent, with fixed dwellings; however, transient villages can occur. Further, the dwellings of a village are fairly close to one another, not scattered broadly over the landscape, as a dispersed settlement.

The village of Rougon in Provence, France

In the past, villages were a usual form of community for societies that practise subsistence agriculture, and also for some non-agricultural societies. In Great Britain, a hamlet earned the right to be called a village when it built a church.[1] In many cultures, towns and cities were few, with only a small proportion of the population living in them. The Industrial Revolution attracted people in larger numbers to work in mills and factories; the concentration of people caused many villages to grow into towns and cities. This also enabled specialization of labor and crafts, and development of many trades. The trend of urbanization continues, though not always in connection with industrialisation. Villages have been eclipsed in importance as units of human society and settlement.

An alpine village in the Lötschental Valley, Switzerland

Traditional villages

Although many patterns of village life have existed, the typical village was small, consisting of perhaps 5 to 30 families. Homes were situated together for sociability and defence, and land surrounding the living quarters was farmed. Traditional fishing villages were based on artisan fishing and located adjacent to fishing grounds.

Berber village in Ourika valley, High Atlas, Morocco.

South Asia

India "The soul of India lives in its villages", declared M. K. Gandhi[2] at the beginning of 20th century. According to the 2011 census of India, 68.84% of Indians (around 833.1 million people) live in 640,867 different villages.[3] The size of these villages varies considerably. 236,004 Indian villages have a population less than 500, while 3,976 villages have a population of 10,000+. Most of the villages have their own temple, mosque or church depending on the local religious following.

A North Indian village in Rajasthan, India

A South Indian Village in Andhra Pradesh, India

East Asia

People's Republic of China In mainland China, villages are divisions under township Zh:or town Zh:.

Republic of China (Taiwan) In the Republic of China (Taiwan), villages are divisions under townships or county-controlled cities. The village is called a *tsuen* or *cūn* () under a rural township () and a *li* () under an urban township () or a county-controlled city. See also Li (unit).

Shirakawa-gō, Gifu Japan.

Japan

South Korea

Southeast Asia

Thailand

Brunei, Indonesia, Malaysia and Singapore

In Indonesia, depending on the principles they are administered, villages are called *desa* or *kelurahan.* A *desa* (a term that derives from a Sanskrit word meaning "country" that is found in a name such as "Bangladesh") is administered according to traditions and customary law (*adat*), while a *kelurahan* is administered along more "modern" principles. *Desa* are generally located in rural areas while *kelurahan* are generally urban subdivisions. A village head is respectively called *kepala desa* or *lurah.* Both are elected by the local community. A *desa* or *kelurahan* is itself the subdivision of a *kecamatan* (district), in turn the subdivision of a *kabupaten* (regency).

The *nagari* of Pariangan, West Sumatra.

The same general concept applies all over Indonesia. How ever, there is some variation among the vast numbers of Austronesian ethnic groups. For instance, in Bali villages have been created by grouping traditional hamlets or *banjar*, which constitute the basis of Balinese social life. In the Minangkabau country in West Sumatra province traditional villages are called *nagari* (a term deriving from another Sanskrit word meaning "city", which can be found in a name like "Srinagar"). In some areas such as Tanah Toraja, elders take turns watching over the village at a command post. As a general rule, *desa* and *kelurahan* are groupings of hamlets (*kampung* in Indonesian, *dusun* in the Javanese language, *banjar* in Bali).

In Malaysia, the term *kampung* (sometimes spelling *kampong*) in the English language has been defined specifically as "a Malay hamlet or village in a Malay-speaking country".[4] In other words, a ***kampung*** is defined today as a village in Brunei, Indonesia, Singapore, and Malaysia. In Malaysia, a *kampung* is determined as a locality with 10,000 or fewer people. Since historical times, every Malay village came under the leadership of a *penghulu* (village chief), who has the power to hear civil matters in his village (see Courts of Malaysia for more details). A Malay village typically contains a *"masjid"* (mosque) or *"surau"* (Muslim chapel), paddy fields and Malay houses on stilts. Malay and Indonesian villagers practice the culture of helping one another as a community, which is better known as "joint bearing of burdens" (*gotong royong*),[5] as well as being family-oriented (especially the concept of respecting one's family [particularly the parents and elders]), courtesy and believing in God (*"Tuhan"*) as paramount to everything else. It is common to see a cemetery near the mosque, as all Muslims in the Malay or Indonesian village want to be prayed for, and to receive Allah's blessings in the afterlife. While in Sarawak and East Kalimantan, some villages are called 'long', primarily inhabited by the Orang Ulu.

Singapore also follows the Malaysian *kampung*. However, there are only a few *kampung* villages remaining, mostly on islands surrounding Singapore such as Pulau Ubin. In the past, there was many *kampung* villages in Singapore but now there aren't many on the mainland.

The term "kampung", sometimes spelled "kampong" is one of many Malay words to have entered common usage in Malaysia and Singapore. Locally, the term is frequently used to refer to one's hometown.

Philippines

In urban areas of the Philippines, the term "village" most commonly refers to private subdivisions, especially gated communities. These villages emerged in the mid-20th century and were initially the domain of elite urban dwellers. Those are common in major cities in the country and their residents have a wide range of income levels. Such villages may or may not correspond to administrative units (usually barangays) and/or be privately administered. Barangays more correspond to the villages of old times, and the chairman (formerly a village datu) now settles administrative, intrapersonal, and political matters or polices the village, though with much less authority and respect than in Indonesia or Malaysia.

Vietnam

Village, or "làng", is a basis of Vietnam society. Vietnam's village is the typical symbol of Asian agricultural production. Vietnam's village typically contains: a village gate, "lũy tre" (bamboo hedges), "đình làng" (communal house) where "thành hoàng" (tutelary god) is worshiped, a common well, "đồng lúa" (rice field), "chùa" (temple) and houses of all families in the village. All the people in Vietnam's villages usually have a blood relationship. They are farmers who grow rice and have the same traditional handicraft. Vietnam's villages have an important role in society (Vietnamese saying: "Custom rules the law" -"Phép vua thua lệ làng" [literally: the king's law yields to village customs]). Everyone in Vietnam wants to be buried in their village when they die.

Central and Eastern Europe

Slavic countries

Selo (Cyrillic: село; Polish: *wieś, sioło*) is a Slavic word meaning "village" in Bosnia and Herzegovina, Bulgaria, Croatia, Macedonia, Russia, Serbia, and Ukraine. For example there are numerous *sela* (plural of *selo*) called Novo Selo in Bulgaria, Croatia, Montenegro and others in Serbia, and Macedonia. In Slovenia, the word *selo* is used for very small villages (less than a thousand people) and in dialects; the Slovene word *vas* is used all over Slovenia.

Bulgaria

Kovachevitsa, a village in southern Bulgaria

In Bulgaria, the different types of *Sela* vary from a small selo of 5 to 30 families to one of several thousand people. According to a 2002 census, in that year there were 2,385,000 Bulgarian citizens living in settlements classified as *villages*.[6] A 2004 Human Settlement Profile on Bulgaria[7] conducted by the United Nations Department of Economic and Social Affairs stated that:

> The most intensive is the migration "city – city". Approximately 46% of all migrated people have changed their residence from one city to another. The share of the migration processes "village – city" is significantly less – 23% and "city – village" – 20%. The migration "village – village" in 2002 is 11%.[6]

It also stated that

> the state of the environment in the small towns and villages is good apart from the low level of infrastructure.[6]

In Bulgaria, it is becoming popular to visit villages for the atmosphere, culture, crafts, hospitality of the people and the surrounding nature. This is called *selski tourism* (Bulgarian: селски туризъм), meaning "village tourism".

Russia

The village near Lake Baikal, Siberia

In Russia, as of the 2010 Census, 26.3% of the country's population lives in rural localities;[8] down from 26.7% recorded in the 2002 Census.[8] Multiple types of rural localities exist, but the two most common are *derevnya* (деревня) and *selo* (село). Historically, the formal indication of status was religious: a city (*gorod*) had a cathedral, a *selo* had a church, while a *derevnya* had neither.

The lowest administrative unit of the Russian Empire, a *volost*, or its Soviet or modern Russian successor, a *selsoviet*, was typically headquartered in a *selo* and embraced a few neighboring villages.

Between 1926 and 1989, Russia's rural population shrank from 76 million people to 39 million, due to urbanization, collectivization, dekulakization, and the World War II losses, but has nearly stabilized since. During 1930–1937, mass starvation in Russia and other parts of the Soviet Union lead to the death of at least 14.5 million peasants (including 5–7 million in the Holodomor).[9]

Most Russian rural localities have populations of less than 200 people, and the smaller places take the brunt of depopulation: e.g., in 1959, about one half of Russia's rural population lived in villages of fewer than 500 people, while now less than one third does. In the 1960s–1970s, the depopulation of the smaller villages was driven by the central planners' drive to get the farm workers out of smaller, "prospect-less" hamlets and into the collective or state farms' main villages, with more amenities.[10]

Most Russian rural residents are involved in agricultural work, and it is very common for villagers to produce their own food. As prosperous urbanites purchase village houses for their second homes, Russian villages sometimes are transformed into dacha settlements, used mostly for seasonal residence.

The historically Cossack regions of Southern Russia and parts of Ukraine, with their fertile soil and absence of serfdom, had a rather different pattern of settlement from central and northern Russia. While peasants of central Russia lived in a village around the lord's manor, a Cossack family often lived on its own farm, called *khutor*. A number of such *khutors* plus a central village made up the administrative unit with a center in a *stanitsa* (Russian: стани́ца; Ukrainian: станиця, *stanytsia*). Such *stanitsas* often with a few thousand residents, were usually larger than a typical *selo* in central Russia.

The term *aul*/*aal* is used to refer mostly Muslim-populated villages in Caucasus and Idel-Ural, without regard to the number of residents.

Ukraine

The largest Ukrainian village ("selo") Kosmach

In Ukraine, a village, known locally as a *"selo"* (село), is considered the lowest administrative unit. Villages may have an individual administration (*silrada*) or a joint administration, combining two or more villages. Villages may also be under the jurisdiction of a city council (*miskrada*) or town council (*selyshchna rada*) administration.

There is, however, another smaller type of settlement which is designated in Ukrainian as a *selysche* (селище). This type of community is generally referred to in English as a "settlement". In comparison with an urban-type settlement, Ukrainian legislation does not have a concrete definition or a criterion to differentiate such settlements from villages. They represent a type of a small rural locality that might have once been a *khutir*, a fisherman's settlement, or a dacha. They are administered by a *silrada* (council) located in a nearby adjacent village. Sometimes the term *"selysche"* is also used in a more general way to refer to adjacent settlements near a bigger city, including urban-type settlements (*selysche miskoho typu*) and/or villages; however, ambiguity is often avoided in connection with urbanized settlements by referring to them using the three-letter abbreviation *smt* instead.

The *khutir* (хутір) and *stanytsia* (станиця) are not part of the administrative division any longer, primarily due to collectivization. *Khutirs* were very small rural localities consisting of just few housing units and were sort of individual farms. They became really popular during the Stolypin reform in the early 20th century. During the collectivization, however, residents of such settlements were usually declared to be kulaks and had all their property confiscated and distributed to others (nationalized) without any compensation. The *stanitsa* likewise has not survived as an administrative term. The *stanitsa* was a type of a collective community that could include one or more settlements such as villages, *khutirs*, and others. Today, *stanitsa*-type formations have only survived in Kuban (Russian Federation) where Ukrainians were resettled during the time of the Russian Empire.

Western & Southern Europe

United Kingdom

A village in the UK is a compact settlement of houses, smaller in size than a town, and generally based on agriculture or, in some areas, mining (such as Ouston, County Durham), quarrying or sea fishing.

The major factors in the type of settlement are location of water sources, organisation of agriculture and landholding, and likelihood of flooding. For example, in areas such as the Lincolnshire Wolds, the villages are often found along the spring line halfway down the hillsides, and originate as spring line settlements, with the original open field systems around the village. In northern Scotland, most villages are planned to a grid pattern located on or close to major roads, whereas in areas such as the Forest of Arden, woodland clearances produced small hamlets around village greens.[11] [12]

The main street of the village of Castle Combe, Wiltshire, England.

Some villages have disappeared (for example, deserted medieval villages), sometimes leaving behind a church or manor house and sometimes nothing but bumps in the fields.Some show archaeological evidence of settlement at three or four different layers, each distinct from the previous one. Clearances may have been to accommodate sheep or game estates, or enclosure, or may have resulted from depopulation, such as after the Black Death or following a move of the inhabitants to more prosperous districts. Other villages have grown and merged and often form hubs within the general mass of suburbia — such as Hampstead, London and Didsbury in Manchester. Many villages are now predominantly dormitory locations and have suffered the loss of shops, churches and other facilities.

For many British people, the village represents an ideal of Great Britain. Seen as being far from the bustle of modern life, it is represented as quiet and harmonious, if a little inward-looking. This concept of an unspoilt Arcadia is present in many popular representations of the village such as the radio serial *The Archers* or the best kept village competitions.[13]

Bisley, Gloucestershire, a village in the Cotswolds

Many villages in South Yorkshire, North Nottinghamshire, North East Derbyshire, County Durham, South Wales and Northumberland are known as pit villages. These (such as Murton, County Durham) grew from hamlets when the sinking of a colliery in the early 20th century resulted in a rapid growth in their population and the colliery owners built new housing, shops, pubs and churches. Some pit villages outgrew nearby towns by area and population; for example, Rossington in South Yorkshire came to have over four times more people than the nearby town of Bawtry. Some pit villages grew to become towns; for example, Maltby in South Yorkshire grew from 600 people in the 19th century[14] to over 17,000 in 2007.[15] Maltby was constructed under the auspices of the Sheepbridge Coal and Iron Company and included ample open spaces and provision for gardens.[16]

In the UK, the main historical distinction between a hamlet and a village was that the latter had a church,[1] and so usually was the centre of worship for an ecclesiastical parish. However, some civil parishes may contain more than one village. The typical village had a pub or inn, shops, and a blacksmith. But many of these facilities are now gone, and many villages are dormitories for commuters. The population of such settlements ranges from a few hundred people to around five thousand. A village is distinguished from a town in that:

- A village should not have a regular agricultural market, although today such markets are uncommon even in settlements which clearly are towns.
- A village does not have a town hall nor a mayor.
- If a village is the principal settlement of a civil parish, then any administrative body that administers it at parish level should be called a parish council or parish meeting, and not a town council or city council. However, some civil parishes have no functioning parish, town, or city council nor a functioning parish meeting. In Wales, where the equivalent of an English civil parish is called a Community, the body that administers it is called a Community Council. However, larger councils may elect to call themselves town councils.[17] Unlike Wales, Scottish community councils have no statutory powers.[18]
- There should be a clear green belt or open fields, as, for example, seen on aerial maps for Ouston surrounding its parish[19] borders. However this may not be applicable to urbanised villages: although these may not considered to be villages, they are often widely referred to as being so; an example of this is Horsforth in Leeds.

France

Same general definition as in the UK.

An independent association named *Les Plus Beaux Villages de France*, was created in 1982 to promote assets of small and picturesque French villages of quality heritage. As of 2008, 152 villages in France have been listed in "The Most Beautiful Villages of France".

Saint-Cirq-Lapopie (Lot) is one of "The Most Beautiful Villages in France".

Spain

Spain has plenty of little villages around its territory. The concept of village and country life is really present and usual in the North of the country (Atlantic area), especially in Galicia where villages are similar to English ones.

South of Barcelona is Spain's most romantic Mediterranean beach town, with a 2.5 km-long (1 1/2-mile) sandy beach and a promenade studded with flowers and palm trees. Sitges is a town with a rich connection to art; Picasso and Dalí both spent time here.[20] Mérida is an important Roman town with great tapas. Barcena Mayor (Cantabria) has houses that date back to the sixth century with simple two floors constructions and rectangular form. Salamanca, an ancient Celtic town, is also a Renaissance city with striking architecture. Its sandstones buildings have a beautiful lustre giving the city the nickname, La Ciudad Dorado.[21]

Morella, Castellón is a medieval village located in the region of "Comunitat Valenciana" with huge castles with a rich renaissance history.[22] Rogueira pasturelands is one of the great ecological jewels of Galicia. Rivers, pools and springs abound in this verdant forest, as do underground water caves and caverns with a prehistoric past.[23] San Marti Vell is a charming small village well known for its Gothic spire. La Bisbal should be next on the list. The town is worth visiting for its Main Square and the castle. This Romanesque castle is situated in the middle of the town, giving it a romantic look. There is also Palafrugell, Palau-sator, Sant Julia and Sant Feliu de Boada. They are all very important because of their medieval patrimony. Castelló d'Empúries has 13th century Gothic churches. Angles also possesses outstanding medieval constructions throughout its village.[24] All villages have a church or hermitage.

Portugal

Villages are more usual in the northern and central regions and in the Alentejo. Most of them have a church and a "Casa do Povo" (people's house), where the village's summer **romarias** or religious festivities are usually held. Summer is also when many villages are host to a range of folk festivals and fairs, taking advantage of the fact that many of the locals who reside abroad tend to come back to their native village for the holidays.

Netherlands

In the flood prone districts of the Netherlands, villages were traditionally built on low man-made hills called terps before the introduction of regional dyke-systems. In modern days, the term *dorp* (lit. "village") is usually applied to settlements no larger than 20,000, though there's no official law regarding status of settlements in the Netherlands.

Middle East

Lebanon

The main square of Saifi Village in Centre Ville, Beirut, Lebanon

Like France, villages in Lebanon are usually located in remote mountainous areas. The majority of villages in Lebanon retain their Aramaic names or are derivative of the Aramaic names, and this is because Aramaic was still in use in Mount Lebanon up to the 18th century.[25]

Many of the Lebanese villages are a part of districts, these districts are known as "kadaa" which includes the districts of Baabda (Baabda), Aley (Aley), Matn (Jdeideh), Keserwan (Jounieh), Chouf (Beiteddine), Jbeil (Byblos), Tripoli (Tripoli), Zgharta (Zgharta / Ehden), Bsharri (Bsharri), Batroun (Batroun), Koura (Amioun), Miniyeh-Danniyeh (Minyeh / Sir Ed-Danniyeh), Zahle (Zahle), Rashaya (Rashaya), Western Beqaa (Jebjennine / Saghbine), Sidon (Sidon), Jezzine (Jezzine), Tyre (Tyre), Nabatiyeh (Nabatiyeh), Marjeyoun (Marjeyoun), Hasbaya (Hasbaya), Bint Jbeil (Bint Jbeil), Baalbek (Baalbek), and Hermel (Hermel).

The district of Danniyeh consists of thirty six small villages, which includes Almrah, Kfirchlan, Kfirhbab, Hakel al Azimah, Siir, Bakhoun, Miryata, Assoun, Sfiiri, Kharnoub, Katteen, Kfirhabou, Zghartegrein, Ein Qibil.

Danniyeh (known also as Addinniyeh, Al Dinniyeh, Al Danniyeh, Arabic: سير الضنية) is a region located in Miniyeh-Danniyeh District in the North Governorate of Lebanon. The region lies east of Tripoli, extends north as far as Akkar District, south to Bsharri District and Zgharta District and as far east as Baalbek and Hermel. Dinniyeh has an excellent ecological environment filled with woodlands, orchards and groves. Several villages are located in this mountainous area, the largest town being Sir Al Dinniyeh.

An example of a typical mountainous Lebanese village in Dannieh would be Hakel al Azimah which is a small village that belongs to the district of Danniyeh, situated between Bakhoun and Assoun's boundaries. It is in the centre of the valleys that lie between the Arbeen Mountains and the Khanzouh.

Syria

General view from Al-Annaze village, near Tartus, Syria

Syria contains a large number of villages that vary in size and importance, including the ancient, historical and religious villages, such as Ma'loula, Sednaya, and Brad (Mar Maroun's time). The diversity of the Syrian environments creates significant differences between the Syrian villages in terms of the economic activity and the method of adoption. Villages in the south of Syria (Hauran, Jabal al-Druze), the north-east (the Syrian island) and the Orontes River basin depend mostly on agriculture, mainly grain, vegetables and fruits. Villages in the region of Damascus and Aleppo depend on trading. Some other villages, such as Marmarita depend heavily on tourist activity.

Mediterranean cities in Syria, such as Tartus and Latakia have similar types of villages. Mainly, villages were built in very good sites which had the fundamentals of the rural life, like water. An example of a Mediterranean Syrian village in Tartus would be al-Annaze, which is a small village that belongs to the area of al-Sauda. The area of al-Sauda is called a nahiya, which is a subdistrict.

Australasia and Oceania

The village of Puamau on Hiva Oa, Marquesas Islands, French Polynesia

Pacific Islands Communities on pacific islands were historically called villages by English speakers who traveled and settled in the area. Some communities such as several Villages of Guam continue to be called villages despite having large populations that can exceed 40,000 residents.

New Zealand The traditional Māori village was the pā, a fortified hill-top settlement. Tree-fern logs and flax were the main building materials. As in Australia (see below) the term is now used mainly in respect of shopping or other planned areas.

Australia The term village often is used in reference to small planned communities such as retirement communities or shopping districts, and tourist areas such as ski resorts. Small rural communities are usually known as townships. Larger settlements are known as towns.

South America

Argentina Usually set in remote mountainous areas, some also cater to winter sports and/or tourism, see: Uspallata, La Cumbrecita, Villa Traful and La Cumbre

North America

Canada

A Newfoundland fishing village

United States

Incorporated villages

A church in Newfane, Vermont

In twenty[26] U.S. states, the term "village" refers to a specific form of incorporated municipal government, similar to a city but with less authority and geographic scope. However, this is a generality; in many states, there are villages that are an order of magnitude larger than the smallest cities in the state. The distinction is not necessarily based on population, but on the relative powers granted to the different types of municipalities and correspondingly, different obligations to provide specific services to residents.

In some states such as New York, Wisconsin, or Michigan, a village is an incorporated municipality, usually, but not always, within a single town or civil township. Residents pay taxes to the village and town or township and may vote in elections for both as well. In some cases, the village may be coterminous with the town or township. There are also many villages which span the boundaries of more than one town or township, and some villages may even straddle county borders.

There is no limit to the population of a village in New York; Hempstead, the largest village in the state, has 55,000 residents, making it more populous than some of the state's cities. However, villages in the state may not exceed five square miles (13 km²) in area.

In the state of Wisconsin, a village is always legally separate from the towns that it has been incorporated from. The largest village is Menomonee Falls, which has over 32,000 residents.

Michigan and Illinois also have no set population limit for villages and there are many villages that are larger than cities in those states. The village of Arlington Heights, IL had 75,101 residents as of the 2010 census.

Villages in Ohio are often legally part of the township from which they were incorporated, although exceptions such as Hiram exist, in which the village is separate from the township.[27] They have no area limitations, but become cities if they grow a population of more than 5,000.[28]

In Maryland, a locality designated "Village of ..." may be either an incorporated town or a special tax district.[29] An example of the latter is the Village of Friendship Heights.

In states that have New England towns, a "village" is a center of population or trade, including the town center, in an otherwise sparsely-developed town or city — for instance, the village of Hyannis in the city of the Barnstable, Massachusetts.

Unincorporated villages

In many states, the term "village" is used to refer to a relatively small unincorporated community, similar to a hamlet in New York state. This informal usage may be found even in states that have villages as an incorporated municipality, although such usage might be considered incorrect and confusing.

Africa

Nigeria

Villages in Nigeria vary significantly because of cultural and geographical differences.

A village in Kaita, NigeriaKaita Nigeria

Northern Nigeria

In the North, villages were under traditional rulers long before the Jihad of Shaikh Uthman Bin Fodio and after the Holy War. At that time Traditional rulers used to have absolute power in their administrative regions. After Dan Fodio's Jihad in 1804,[30] political structure of the North became Islamic where emirs were the political, administrative and spiritual leaders of their people. These emirs appointed a number of people to assisted them in running the administration and that included villages. [31]

Every Hausa village was reigned by Magaji (Village head) who was answerable to his Hakimi (mayor) at town level. The Magaji also had his cabinet who assisted him rule his village efficiently, among whom was Mai-Unguwa (Ward Head). [32]

With the creation of Native Authority in Nigerian provinces, the autocratic power of village heads along with all other traditional rulers was subdued hence they ruled 'under the guidance of colonial officials'.[33]

Even though the constitution of the Federal Republic of Nigeria has not recognised the functions of traditional rulers, they still command respect in their villages[33] and political office holders liaise with them almost every time to reach people.

In Hausa language, village is called **ƙauye** and every local government area is made up of several small and large **ƙauyuka** (villages). For instance, Girka is a village in Kaita town in Katsina state in Nigeria. They have mud houses with thatched roofing though, like in most of villages in the North, zinc roofing is becoming a common sight.

Still in many villages in the North, people do not have access to portable water. So they fetch water from ponds and streams. Others are lucky to have wells within a walking distance. Women rush in the morning to fetch water in their clay pots from wells, boreholes and streams. However, government is now providing them with water bore holes.[34]

Electricity and GSM network are reaching more and more villages in the North almost everyday. So bad feeder roads may lead to remote villages with electricity and unstable GSM network.[35]

Southern Nigeria

Village dwellers in the Southeastern region lived separately in 'clusters of huts belonging to the patrilinage'.[36] As the rainforest region is dominated by Igbo speaking people, the villages are called **ime obodo** (inside town) in Igbo language. A typical large village might have a few thousand persons who shared the same market, meeting place and beliefs.

See also

- Global village
- Linear village
- Village green
- Village lock-up
- police village

Settlement types

- Dugout
- Fishing village
- Hamlet
- Microtown

Countries and localities

- Dhani and villages
- Dogon villages
- Hakka architecture
- Ksar
- List of villages in Europe by country
- Pueblo
- Sołectwo (rough equivalent in Poland)
- Ville

Developed environments

- Developed environments
- City
- Exurban

- Megalopolis
- Rural
- Suburban
- Urban area

Footnotes

[1] Dr Greg Stevenson, "What is a Village?" (http://www.bbc.co.uk/history/programmes/restoration/2006/exploring_brit_villages_01.shtml), *Exploring British Villages*, BBC, 2006, accessed 20 October 2009

[2] R.K. Bhatnagar. INDIA'S MEMBERSHIP OF ITER PROJECT (http://www.pibbng.kar.nic.in/feature1.pdf). PRESS INFORMATION BUREAU. GOVERNMENT OF INDIA, BANGALORE

[3] "Indian Census" (http://www.censusindia.gov.in/). Censusindia.gov.in. . Retrieved 2012-04-09.

[4] "Meriam-Webster Online" (http://www.m-w.com/dictionary/kampung). M-w.com. 2007-04-25. . Retrieved 2010-03-28.

[5] Geertz, Clifford. "Local Knowledge: Fact and Law in Comparative Perspective", pp. 167–234 in Geertz *Local Knowledge: Further Essays in Interpretive Anthropology*, NY: Basic Books. 1983.

[6] "Human Settlement Country Profile, Bulgaria (*2004*)" (http://www.un.org/esa/agenda21/natlinfo/countr/bulgaria/Bulgariahumansettlement2003.PDF) (PDF). United Nations Department of Economic and Social Affairs. . Retrieved 2008-11-30.

[7] HUMAN SETTLEMENT COUNTRY PROFILE: BULGARIA (http://www.un.org/esa/agenda21/natlinfo/countr/bulgaria/Bulgariahumansettlement2003.PDF). United Nations (2004)

[8] "Предварительные итоги Всероссийской переписи населения 2010 года[[Category:Articles containing non-English language text (http://www.perepis-2010.ru/results_of_the_census/results-inform.php)] [Preliminary results of the 2010 All-Russian Population Census]"] (in Russian). *Всероссийская перепись населения 2010 года (2010 All-Russia Population Census)*. Federal State Statistics Service. 2011. . Retrieved February 9, 2012.

[9] Robert Conquest (1986) *The Harvest of Sorrow: Soviet Collectivization and the Terror-Famine*. Oxford University Press. ISBN 0-19-505180-7.

[10] "Российское село в демографическом измерении" (*Rural Russia measured demographically*) (http://demoscope.ru/weekly/2006/0253/tema04.php) (**Russian**). This article reports the following census statistics:

Census year	1959	1970	1979	1989	2002
Total number of rural localities in Russia	294,059	216,845	177,047	152,922	155,289
Of them, with population 1 to 10 persons	41,493	25,895	23,855	30,170	47,089
Of them, with population 11 to 200 persons	186,437	132,515	105,112	80,663	68,807

[11] Wild, Martin Trevor (2004). *Village England: a social history of the countryside* (http://books.google.co.uk/books?id=M7AmyuGr5Y8C&printsec=frontcover). I.B.Tauris. p. 12. ISBN 978-1-86064-939-4. .

[12] Taylor, Christopher (1984). *Village and farmstead: A history of rural settlement in England* (http://books.google.co.uk/books?ei=NksHTpXqIo6t8QPE1ozADQ). G. Philip. p. 192. ISBN 978-0-540-01082-0. .

[13] OECD (2011). *OECD Rural Policy Reviews: England, United Kingdom 2011* (http://books.google.co.uk/books?id=bQCGWKfXJNMC&printsec=frontcover&source=gbs_ge_summary_r&cad=0). OECD Publishing. p. 237. .

[14] *The Parliamentary gazetteer of England and Wales* (http://books.google.co.uk/books?id=mxIQAAAAYAAJ&printsec=frontcover). **3**. A. Fullarton & Co.. 1851. p. 344. .

[15] "Maltby Ward" (http://www.rotherham.gov.uk/download/553/maltby_ward). Rotherham Metropolitan Borough Council. . Retrieved 2011-06-26.

[16] Baylies, Carolyn Louise (1993). *The history of the Yorkshire miners, 1881–1918* (http://books.google.co.uk/books?id=WEIOAAAAQAAJ&printsec=frontcover&source=gbs_ge_summary_r&cad=0). Routledge. .

[17] "National Statistics" (http://www.statistics.gov.uk/geography/parishes.asp). Statistics.gov.uk. . Retrieved 2010-03-28.

[18] "Portobello Community Council" (http://www.porty.org.uk/council/index.php). Porty.org.uk. . Retrieved 2010-03-28.

[19] "Ouston Parish Council" (http://parishes.durham.gov.uk/ouston/Pages/wherewelive.aspx). durham.gov.uk. .

[20] The Best Small Towns and Villages in Spain at Frommer's (http://www.frommers.com/destinations/spain/0242020855.html). Frommers.com. Retrieved on 2012-05-27.

[21] Spanish Towns and Villages (http://www.tourclare.com/spanishtownsandvillages.php). Tourclare.com. Retrieved on 2012-05-27.

[22] Top 10 Cities and Villages to Visit in Spain (http://www.travelthruspain.com/what-to-do/cities). Travelthruspain.com. Retrieved on 2012-05-27.

[23] Medieval villages in Galicia, Spain: O courel mountains. spain.info in English (http://www.spain.info/no/reportajes/sierra_de_o_courel_naturaleza_y_aldeas_medievales.html). Spain.info. Retrieved on 2012-05-27.

[24] Medieval Towns And Villages In Spain (http://travel.ezinemark.com/medieval-towns-and-villages-in-spain-16bc87fb4dc.html). Travel.ezinemark.com (2010-10-22). Retrieved on 2012-05-27.

[25] "A project proposal" (http://almashriq.hiof.no/lebanon/400/410/412/elies_project/glimse_of_yesterday.html). Almashriq.hiof.no. . Retrieved 2010-03-28.

[26] "Village" (http://www.websters-online-dictionary.org/definition/english/vi/village.html#Definitions). Websters-online-dictionary.org. . Retrieved 2010-03-28.

[27] "Detailed map of Ohio" (http://www2.census.gov/geo/maps/general_ref/cousub_outline/cen2k_pgsz/oh_cosub.pdf) (PDF). United States Census Bureau. 2000. . Retrieved 2010-03-28.

[28] "Ohio Revised Code Section 703.01(A)" (http://codes.ohio.gov/orc/703.01). . Retrieved 2010-03-28.

[29] 2002 Census of Governments, Individual State Descriptions (http://www.census.gov/prod/2005pubs/gc021x2.pdf) (PDF)

[30] Britannica Encylopedia **The New Encyclopaedia Britannica** Vol. 6, 15th Edition. isbn 0-85229-961-3. Page 763

[31] Dr. Sani Abubakar Lugga. **The Great Province** published by Lugga Press Gidan Lugga, Kofar Marusa Road, Katsina Nigeria. ISBN 978-2105- 48 - 1. Page 43

[32] Dr. Sani Abubakar Lugga. **The Great Province** published by Lugga Press Gidan Lugga, Kofar Marusa Road, Katsina Nigeria. ISBN 978-2105- 48 - 1. Page 63

[33] A Johnson Ugoji Anyaele. **Comprehensive Government** A Johnson Publishers LTD. Surulere, Lagos. ISBN 978-2799-49-1. Page 123

[34] Katsina state government spends N4.2 billion on water supply project each year. http://www.peoplesdaily-online.com/index.php/news/news/special-report/7632-how-katsina-state-is-doing-so-much-with-so-little

[35] Nigerian Operator Expands Coverage. http://www.cellular-news.com/story/16837.php

[36] Villages in Igbo land prior to the colonial era. http://www.igboguide.org/HT-chapter10.htm

External links

- Types of villages (anthropogenic biomes) (http://www.ecotope.org/anthromes/v1/guide/villages/)

Article Sources and Contributors

Rogóźno, Łask County *Source*: http://en.wikipedia.org/w/index.php?title=Rog%C3%B3%C5%BAno%2C_%C5%81ask_County *Contributors*: D6

Łódź *Source*: http://en.wikipedia.org/w/index.php?title=%C5%81%C3%B3d%C5%BA *Contributors*: *drew, 123Sister, A bit iffy, A. B., Aaronbrick, Abramakabreska, Ada Kruk, AdjustShift, Aervanath, Aeusoes1, Alaniaris, AlexTiefling, Allisonckistler, Allreet, Amnon s, Andersao, Andres, Andrew Hampe, Ani td, AshleyMorton, Austrian, Avicennasis, B0sh, Badyl4325, Bahar101, Balcer, BalkanFever, Beaumont, Biblicalscholar99, BigrTex, Biladek, Biruitorul, Black Stripe, Blewenstein, Bobanni, Bogomolov.PL, Boyar, Brisvegas, Bulwersator, Bunnyhop11, Burzmali, BuzzWoof, CJLL Wright, CLW, CWO, Caius2ga, Calton, Carelesskiss, Cautious, Ceoil, Cestina2008, Cezary Okupski, Cfs198, Chanheigeorge, Charles01, Chris G, Chris j wood, CiaPan, CieloEstrellado, ClaudiusMaius, Cocytus, CommonsDelinker, Crazycomputers, CredoFromStart, Cst17, Curps, Cyanidesandwich, Cyrillic, Dahn, Danny, Darwinek, DerBorg, DerHexer, Dezidor, Dohn joe, Dr Sameer Ayyoub, Dr. Dan, Dysprosia, Dzied Bulbash, EBL, Edgar181, Edinborgarstefan, Egor9, El C, Elk Salmon, Elonka, Emax, Emil Kwas, EoGuy, Epeefleche, Escottf, Eugene van der Pijll, Everard Proudfoot, Ewulp, Filelakeshoe, Filii Henoch, Football in Poland, Fratrep, Fujicolor, GCarty, GagHalfrunt, Gaspar van der Sar, Gauss, Gene Nygaard, Georgecmu, Gerry Lynch, Ghirlandajo, Gilabrand, Girolamo Savonarola, Gnesener1900, Green Giant, Gunnar Larsson, Gzornenplatz, Hajor, Halibutt, HerkusMonte, Historicist, Hiuppo, Hob, Hturko, HuBar, Ian Pitchford, Imperium Europeum, In ictu oculi, IronGargoyle, Irpen, J04n, Jack Merridew, James086, JamieS93, Janggeom, Jaxdelaguerre, Jeronimo, Jhendin, Jim, Jimp, Jj137, John Quincy Adding Machine, JohnArmagh, JonnyLightning, Joseph Solis in Australia, Jpbowen, Jumbuck, Kajzderski, Karol Langner, Kasnie, Kelisi, Kevinhksouth, Khatru2, Khoikhoi, Khrystene, Kitauga, Kjoonlee, Koavf, Kotniski, Kpalion, Kpjas, Kronidos, Kruanne, Kubeu, Kwamikagami, L Kensington, LMB, La Parka Your Car, Lightmouse, Lil downloader, Lilac Soul, LittleWink, Livajo, Lotje, Lurlock, Lysy, M Czerwiowski, M.laube, MJCdetroit, MKoltnow, MacTire02, MacedonianBoy, Magioladitis, Malcolm, Malick78, Manneken Pis, Marek69, Mareklug, Mark, Marskell, Mathew5000, Maximus Rex, Mazuria, Miles530, Mirek46, Mogism, Molobo, Moszczynski, Mr. Stradivarius, Msamj, Msginsberg, Muminekkasia, Mutt Lunker, NYCfellow, Naddy, Nakon, Natrajdr, Nazroon, Neutrality, Nick Number, NickelShoe, Nihola, Nikai, NuclearWarfare, Ohconfucius, OldDummy, Olessi, Oobopshark, Ophitke38, OwenBlacker, PamD, Pedant17, Peliti, Penguin, Pernambuko, Petri Krohn, Phlebas, Phyllis1753, Pigsonthewing, Piotrus, Pippirrup, Poeticbent, Polimerek, PolishPoliticians, Polotet, Polska1991, Quarl, Queenmomcat, Rd232, Redaktor, Rekakaka, Rich Farmbrough, Rjwilmsi, Rmhermen, Robertgreer, Robertson-Glasgow, Romuald Wróblewski, Rosh2610, Rozowa owca, SGreen, Sanacja, Sapphire, Saranghae honey, Sardanaphalus, Satori Son, Savidan, Savvo, Schmichaeln, SchuminWeb, Shalom Alechem, Shizhao, Shogun, SidP, Siebe, SiobhanHansa, Sjdunn9, Skalee vel crensh, Sliwers, Smetanahue, Spartan 1200, Spellcast, SpiderMum, Spinning Giraffe, Stepheng3, Stoshik, Sturunner, Susvolans, Sweden555, Szeged, Tdurden78, Tellerman, Template namespace initialisation script, Tpbradbury, Triskaidekafil, Tsuba, Turpster, Tymek, Unschool, Urhixidur, Ventusa, VeryVerily, Vinsentti, Viol8or, Voceditenore, Volunteer Marek, Voyevoda, Vumba, Welsh, Wetman, Wickiss0, Wik, Witkacy, WojPob, Woohookitty, Wurzeller, Wwoods, Xanoxis, XpertPL, Y, YagaBaba, Yurivict, Александър, 502 anonymous edits

Warsaw *Source*: http://en.wikipedia.org/w/index.php?title=Warsaw *Contributors*: 08OceanBeach SD, 1exec1, 32X, 334a, 5-HT8, 52 Pickup, 54gsze4ghz5, A8UDI, Aaroncrick, Adashiel, Adavidb, Adderbak, Addie555, Aditya Mahar, AdjustShift, Adriano 7, Adziura, Aegis Maelstrom, Agnus, Ahoerstemeier, Aitias, Ajh1492, Alan.ca, Alansohn, AlbertR, Aleksandr Grigoryev, AlexanderKaras, Alexandra lb, Allens, Alsandro, Altenmann, Amakuru, Anaxial, Anbu121, Andre Engels, Andreas Hasanai, Andy M. Wang, Angelo De La Paz, Animeguardxox123, Anishpp, Antandrus, Anthony Appleyard, Apparition11, Appleseed, Arcturus, Aresceo, Arminius, ArnoldPettybone, Arsenikk, Attilios, Atwardow, Avala, Averette, Avono, AxelBoldt, Axeldysko, BD2412, Bachrach44, Bacique, Balcer, Baltic Soul, Barry Kent, Bartel1977, Barticus88, Batbayarb, Bbb23, Beetstra, Bejnar, Beland, Berdov, Bestalex, Betterusername, Bg007, Bgmur, Biala Gwiazda, Bigdumbsex, Biglovinb, BigrTex, BillC, Bletch, Bobo192, Bogdan, Bogey97, Boston999, Bpa, Bpeps, Bpiotrek, Bramley, Brandmeister, Brandmeister (old), Brisvegas, BrownHairedGirl, Brzeszczot, Brzeszczykiewicz, BurgererSF, Bushy moustache, Butros, Bwasiluk, Bwood, Bziomek, CJLL Wright, CNN555, CPAScott, CactusWriter, Caius2ga, Can't sleep, clown will eat me, Canadiano, Candbender, Catapult, Cautious, Cdc, Cenarium, Ceriparrymorgan, Cgoodrich 12, ChKa, Chanheigeorge, Charles Matthews, Chepry, ChicXulub, Chowbok, Chris G, Chris j wood, ChrisGualtieri, Ciceronl, Cloudz679, Colonel Mustard, Colonies Chris, ComUSSR, CommonsDelinker, Conscious, Conversion script, Convi, Courcelles, Cpt.schoener, Crystallina, Curps, Cyberchimp, Czyrko, D, DBigXray, DESU TOYS, DK4, DO'Neil, DVD R W, DaQuirin, Dana boomer, Danny, Darius Dhlomo, DarthSmee, Darwinek, Davehi1, David Parker, Dbenbenn, DePiep, Deb, Debresser, Delicious carbuncle, Delirium, Delldot, DemonicElbow, Der Golem, DerBorg, Derbeth, Diceman, Discospinster, Divčinka, Dodsosk, Doopdoop, Dr. Blofeld, Dr. Dan, Drbug, DreamGuy, Druzhnik, Dryazan, Dubhe.sk, Dufourspitze, Duncancumming, Duster555, DuttyYo, EBL, Eagor, Edgar181, Editorofthewiki, Edupolandcom, Eeamoscopolecrushuva, Egor9, Ejdzej, El C, Elekhh, Elockid, Elonka, Elpaw, Emanisafreak, Emax, Enchanter, Enfantsduparadis, Epbr123, Epeefleche, Erdeniss, Eric Bauman, Erolos, Eugene van der Pijll, EuroLine, Ev, Evlekis, Exok, Explicit, Extra999, EyeSerene, FastLizard4, FernandoAngel, Feydey, Fibonacci, Fifelfoo, Flrn, Flurry1985, Foledman, Folks at 137, Football in Poland, Foundert, Fourdee, FrancoGG, Francs2000, Franekwiki, Frankyaraceae, Fransvannes, Fredrik, Freyr, Fujicolor, Future Perfect at Sunrise, Futurebird, GCarty, Gaius Cornelius, Galis, Galoubet, Gareth.randall, Garfield333, Garzo, Gdabski, Gdansk, Gdarin, Gene Nygaard, Ger-man, Ghirlandajo, GiZiBoNG, Gilado, Gilgamesh, Gilgamesh he, Gilliam, Giraffedata, Gnesener1900, Gobeirne, Gogo Dodo, GoingBatty, Goodnightmush, GraemeL, Graham87, Gralo, Green Giant, Grippenn, Ground Zero, Gryffindor, Grzond, Gsklee, Guinness man, Guliolopez, Gustav von Humpelschmumpel, Gzornenplatz, Hackfish, Hadal, Halibutt, HappyCamper, Heikoh, Heine, Heszy, Heyst, Hiotto667, Hiuppo, HkCaGu, Hmains, Hoodinski, Hvn0413, IAnton7, Ian Pitchford, Ianb, IceUnshattered, Ignatius libre, IgorMagic, Iketsi, Iloni Label, Iloveandrea, Imperium Europeum, InfoQultura, Inoruox, Inter-man, Irishguy, Irpen, Ish warsaw, Isidore, J.delanoy, JCarriker, JForget, JHK, JPD, JRWalko, Jacek Kendysz, Jacksav, Jacurek, Jamesofur, JanaMitz, Jane McCann, Janobr, Jazzmaphone, Jd2718, Jheald, Jhendin, Jim Michael, JinJian, Jkirk2012, Jklin, Joeoettinger, John, John Quincy Adding Machine, JohnInDC, Jorunn, Jose77, Jotel, Joyous!, Joytotheworld982, João P. M. Lima, Jpgordon, Julekmanowski, Julian Mendez, Jumbuck, K. Annoyomous, Kadellar, Kaiser von Europa, Kajzderski, Karen Johnson, Karlos87, Karol Langner, Karolczu, Katarzyna, Keegan, Kerry7374, Kevin, Kferden, Khaadimone, Kitch, Kitia, KizzyB, Klik9, Knarf-bz, Knepflerle, Koavf, Kolombus, Kotniski, Kowalmistrz, Koyaanis Qatsi, Kpalion, Kpjas, Kristen Eriksen, Krupo, Ksenon, Kubek15, Kukini, Kuzia, Kwamikagami, LA2, LOTkid, LUCPOL, Lancashire69, Lars Washington, Le Fou, LeManche211, Leandrod, Lear 21, LeszekB, Leuko, Levineps, Lightmouse, LilHelpa, Linnell, Lissajous, LittleOldMe, Logologist, Loosmark, Lotje, Lovetelaviv, LukWarUrs, Lysy, M0RD00R, M2545, M4a, MER-C, MJBurrage, MJCdetroit, MJD86, MMBBTT, MPerel, MaciejKudra, Magioladitis, Mahmudmasri, Mailer diablo, Malcolm, Mamalala, Mantchi, Marcin.sobczyk, Marek Chalecki, Marek69, Mareklug, Marias87, Mariusemillian, Mark83, Markpeak, Martim33, Martinete, Masterpiece2000, Mathiasrex, Matthew Fennell, Mav, Maya, Mayumashu, Mazzelino, Mboverload, McHooty hoot the owl numbah 4!, Mentifisto, Mercury, Meteor2017, Meursault2004, Mfloryan, Michael Hardy, Michael.gierasimiuk, MichaelTinkler, Mieciu K, Mike Hayes, Mike Rosoft, Mikej007, Mikkip, Minimus1, Mintguy, Mirek46, Mohagheghorg, Mokhov, Molobo, Moncrief, Moonriddengirl, MorganaFiolett, Mrg3105, Mulder1982, MusicGeek101, Muslim lo Juheu, Mxn, Myrtone86, Myszodorn, N panni, Naive cynic, Nakon, Nango14, NaturalBornKiller, NatusRoma, NawlinWiki, Neferkare, NellieBly, Nemeng, NeroN BG, Nickst, Nico, Nihil novi, Nk, Norum, NotAnonymous0, Nufy8, Number10a, Numbo3, Nv8200p, Nwinther, Nyh, Oda Mari, Ohconfucius, Oktan, Olessi, Oli Filth, Oly Majoieka, Omgstupidlag, Omicronpersei8, Orestek, Ostap R, OwenBlacker, Oxymoron83, PS, Paluszak, Paolo914, Pashko 2, Patrick, Patrick Nagel, PatrikR, Paulinho28, Pawello071, Pearle, Pebbles1979, Pensionero, PerV, Pernambuko, Petri Krohn, Philipek, Piccolo Modificatore Laborioso, Picus viridis, Pigsonthewing, Piotrus, Pkmink, Plasticup, Plongerstoop, Pmj, Pointillist, Polaco77, Poland rules the area east of Germany, Polandrulez, PolishPoliticians, Poszwa, Powerek38, Profoss, Profsowa, ProhibitOnions, Przepla, Psubrat2000, Pundit, Purz12, Pusior, R'n'B, R9tgokunks, RG2, RSido, RVJ, Rabbigeorge, Radomil, Radzinski, Random2001, Rcsprinter123, Rd232, Readin, Red Winged Duck, Regibox, Remark knights, Rfl, Rhollenton, Rich Farmbrough, RickK, Ricky81682, Rjwilmsi, RobertG, Robertpolska, Robweiller, Romuald Wróblewski, Ronhjones, RoyBoy, Rrburke, Ruhrjung, Russavia, Russian Luxembourger, Ruszewski, Rwerp, SDC, SJP, Sabbut, SaUyDieBoereKomLei, SamEV, Samwb123, Sardanaphalus, Scientizzle, Sciurinæ, Scott Gall, SeanMack, Seaphoto, Sebesta, Seraphimblade, Severo, Sfu, Shalom Alechem, Siebrand, SimonP, SiobhanHansa, Sk8er121, Skurwiel, Sliwers, Smith Jones, Smith2006, Smog, SnapSnap, Snowmanradio, Snoyes, SnrRailways, Soetermans, South Bay, Sp33dyphil, Space Cadet, SpookyMulder, Springeragh, Ssolbergj, Sssmok1, StaticGull, Stickee, Stlvu2, StoM, Storkk, Strawtree, Superkid72, Superplexus, Superzohar, SylwiaS, Syrenab, Szopen, Szpaq, Szwejk, Szy, TCUfan, TShilo12, Tabletop, Taivo, Tandrum, Terry Vaserum, Texmon, The Epopt, The Thing That Should Not Be, The alliance, The wub, TheTimesAreAChanging, Themilstead, Thingg, Thiridaz, Thmars10, Timir2, Timwi, Tobby72, Tomazzz, Tomica, Topory, Tot12, Tpbradbury, TrueHistoryPoland, Tsca, Tsf, Tuomas, Turpster, Tweenk, Twthmoses, Ukanio, UltioUltionis, Ultras GE, Umedard, UnLoCode, VKing, Valentinian, Vecrumba, Vegalabs, VernoWhitney, Verwolff, Violncello, Viper1992, Viragkamil, Visor, Viva-Verdi, VodkaJazz, Volunteer Marek, Vortexrealm, Voyevoda, Vumba, WarsawRob, Wavelength, Wawsrm, Weregerbil, WhisperToMe, White whirlwind, Wikipelli, William Avery, Witkacy, Wlodi, Wmaciej, Wmahan, WojPob, Wojgniew, Woohookitty, Wurzeller, Xiaoyu of Yuxi, Xnuala, Xx236, Xxlittlekiwixx, Yomomo1, Yone Fernandes, Zachtaylor12, Zarbi1, Ziemniak12314, Zimbabweed, Zoidbergness, Zsero, ਰਜੇਨ੍ਦ੍ਰ ਸਿੰਘ, 1716 anonymous edits

Coat of arms *Source*: http://en.wikipedia.org/w/index.php?title=Coat_of_arms *Contributors*: 123aidan, 1exec1, 28421u2232nfenfcenc, A More Perfect Onion, A little insignificant, A.zstephens, ABF, Addshore, Adhyatma3010, Aervanath, Agentscott00, Ahoerstemeier, Ahuskay, Aka042, Alansohn, Alexius08, Amateur55, Amjad Abdullah, Andre Engels, Angela, AnonMoos, Anonymous from the 21st century, Antalagy, Apeloverage, Arab Hafez, Araignee, ArnoldReinhold, Artyom, Arwel Parry, Asbestos, Ashish Bakshi, Astrotrain, Attilios, BIL, Bakabaka, BanyanTree, Barneca, Basil Bernar, Beetstra, Bennylin, Betacommand, Bewareofdog, Big Adamsky, BilCat, Bissinger, Blauendorn, Bobo192, BolinhasFofas, Borameer, Brandon5485, Brian0918, Brianga, BryanG, Bubbha, C9Rav9c, CALR, CTZMSC3, Cahk, CambridgeBayWeather, Can't sleep, clown will eat me, Capricorn42, CardinalDan, Care, CaribDigita, Carlsmith, Celtus, Centauri, Charruyego, Checkpoint, Christian75, Chwyatt, Circeus, Codex Sinaiticus, CommonsDelinker, Cordless Larry, Correogsk, Craigy144, CristianChirita, Crzrussian, Curb Chain, Custodiansoftime, Czar Brodie, DARTH SIDIOUS 2, DTRY, Danbarnesdavies, Danceswithzerglings, Dancingwombatsrule, Danrok, Darcij, Darth Panda, David Latapie, David Shay, David de Cooman, Dawn Bard, Dbachmann, Dcandeto, Debouncer, Decora, Delirium, Demiurge, Derek Andrews, DeusMP, Dfflick319, Digger845, Discospinster, Dlae, Dogears, Doops, Dr pda, DrHollisCollier, Dthomsen8, Duffman, Dysepsion, ESkog, Eliz81, Ellmist, EncycloPetey, Enviroboy, Enzo Aquarius, Epbr123, Equendil, Eric-Wester, Euchiasmus, Eulyzes, Evadb, Evertype, Everyking, Evil saltine, Evrik, Ezeu, Fanatix, Fengxianjin, FiggyBee, Finavon, Fiyinite7, Flausy, Foxcommunication, FrancoGG, Freakofnurture, FrederickSommer, Freederick, Fritzpoll, Fvasconcellos, Gakusha, Galoubet, Galwhaa, Garryq, GearedBull, Generalseven, Geniac, George.moug, Gilliam, Gimboid13, Gimmetrow, Gire 3pich2005, Glane23, Glenlarson, Gogo Dodo, Gorovich, Gregmatthews, Group29, Grutness, Gsp8181, Gtstricky, Gustav von Humpelschmumpel, Gustavo Szwedowski de Korwin, Hadžija, Halmstad, HamburgerRadio, Hawjam, Heraldics, Herostratus, History2007, Hmains, Hoss789, Hu12, Huntscorpio, INkubusse, IRP, Iaberis, Iamunknown, Igoldste, Imagecrafting, ImperatorExercitus, Inka 888, Innerche, Ipankonin, Iridescent, Islescape, J.delanoy, J36miles, JFreeman, Jackol, JackyR, Jake Wartenberg, JamesBWatson, Jaykjellybean, Jdforrester, JerryFriedman, Jhamez84, Jiang, Jjojj, Jncraton, John, John254, JohnCD, Johnathan Tsui, Jonjames1986, Joseph Solis in Australia, José Luís Ávila Silveira, Jusdafax, Jyusin, Kahner5150, Kaihsu, Kanguole, Kansoku, Karmosin, Keilana, Kelisi, Khoikhoi, King of Hearts, KingNewbs, Kintetsubuffalo, Kiore, Kittybrewster, Kookyunii, Kool Lat'n SD, Kostisl, Kovacho, Kpgh99, Krylonblue83, Kylu, Kyuko, Lacrimosus, Lahiru k, Lambiam, Lammoth, Larry1972, LeaveSleaves, LiDaobing, LibLord, Linkboyz, Lobsterthermidor, Logan, Lord Brantus, Lord Fitzwarin, Loren.wilton, Lradrama, LtNOWIS, M.K, MER-C, MacGonagall, Madhero88, Magicuser3, Magioladitis, Magog the Ogre, Man vyi, Marcelivan, Mark, Mark91, Martykoch, MathHisSci, Mauls, Maycoll F. Vieira, Mb1000, Mellisa Anthony Jones, Mhking, Mich Taylor,

Michael Hardy, Michaelsanders, Midgrid, Mike Rosoft, Milford212, Mingwangx, Mon Vier, Mormegil, Mothmolevna, Mouser, Moyogo, Munboy, Muriel Gottrop, Mushroom, Mygerardromance, Nadav1, Nakon, Nameneko, Neurolysis, NewEnglandYankee, Nilfanion, Nivix, Nkcs, Noropdoropi, Nsaa, Ntsimp, OHCLOUDSUNFOLD, Octane, Orange Suede Sofa, Owned Souls, Oxymoron83, PBS, PDH, Palnatoke, Paul Barlow, Paul venter, Pdcook, Pe-Jo, Persian Poet Gal, Petri Krohn, PhJ, Phaedriel, Philip Trueman, Pieneman1, Pinethicket, Pinkadelica, Pkhg199, PleaseStand, Pmanderson, Pne, Poco a poco, Pojanji, Pokrajac, Porqin, Possum, Prothonotar, Puddhe, Quite, Qxz, Qyd, RandomStringOfCharacters, RapidR, Razakel19, Rdsmith4, Remuel, Rettetast, Reywas92, Rhurtadon, Ricardocolombia, RichardGonzalesV, Rjstott, Robchurch, Rockhall, Ronhjones, Roux, Royboycrashfan, S, SChockey97, SPhotographer, Sadalmelik, Saga City, Salvadorjo, Samuella99, Sbwoodside, Schockey14, SeanR70, Sephiroth storm, Shanel, Silverhorse, SimonP, Simonmetcalf, Sinaloa, Sinhaempire, Sionus, Siroxo, Sistan, Skater, Skull 'n' Femurs, Sky Harbor, SmallPotatoes, Smalljim, Smoon17, Snowmanradio, Snowolf, Sotakeit, SowelBlack, Spencer, Sputnikcccp, Smec, Stansult, Studioce, SupermanReturns, SusikMkr, Sweetloulou, TJ Spyke, Tabletop, Tamfang, Tauex, Tckma, Techman224, Tedder, Teo64x, Terot, Terra Xin, TerraHikaru, Thakurji, The Thing That Should Not Be, Theone00, Thingg, This user has left wikipedia, Thisbites, Thisisborin9, Thomas Blomberg, Thumperward, ThuranX, Tide rolls, Tilstock, Tirthajyoti, Tmopkisn, Tom.k, Tone, Tozzano, Trusilver, Twinxor, TylerDurden1963, Tyrenius, Uhai, UserDoe, Utcursch, VAwebteam, Valentinian, Vanished User 1004, Vasilcho, Vincent Liu, Vrenator, Vuo, Vzbs34, WJBscribe, Wales, Wayne Slam, Whistlesgowhoo, Whpq, Wikipedian2, Wilhelm meis, Witan, Woohookitty, Writerguy, Writtenright, Xanderliptak, Xn4, Yamakiri, Yarp, Yopie, Yworo, Zalgo, Zalhael, Zanhe, Zarthon, Zaxo1994, , , 910 anonymous edits

Canting arms *Source*: http://en.wikipedia.org/w/index.php?title=Canting_arms *Contributors*: Alci12, AlmostReadytoFly, Andres, AnonMoos, AnotherOnymous, Anthony Appleyard, Brendandh, Calliopejen1, Circeus, Dbachmann, Denelson83, Donreed, E.G., EncycloPetey, Fastifex, Fishal, Gazilion, Grutness, Heralder, Icairns, Kelisi, Kimon, Kurzhaar, Lhademmor, Liftarn, MPorciusCato, Mailer diablo, Martin23230, MattGiuca, Ndanielm, Pitke, Spellcast, SpikeToronto, Tamfang, Toby Bartels, Wilhelm meis, 31 anonymous edits

Łódź Voivodeship *Source*: http://en.wikipedia.org/w/index.php?title=%C5%81%C3%B3d%C5%BA_Voivodeship *Contributors*: AN(Ger), Adam78, Anten, Appleseed, Backspace, Balcer, Biruitorul, Bobanni, Bonás, Bushy moustache, Caius2ga, Chanheigeorge, CharlotteWebb, Chris G, CommonsDelinker, Docu, Dominus Vobisdu, Ecgrimm, Elapsed, Elonka, Emax, Gaspar van der Sar, GvmBG, Halibutt, Hmains, JeffWexler, Jvano, Kelisi, Kotniski, LittleDan, Logologist, Lotje, MER-C, MaGioZal, MacTire02, Marek69, Mazur, Micga, Nihil novi, Njupas, Orioane, Piotrus, Plastikspork, PolishPoliticians, ProhibitOnions, Radzinski, Revth, Sabbut, Sanya, Sapphire, Sicherlich, Slawojarek, Spacejam2, SpiderMum, Sylwia Ufnalska, Template namespace initialisation script, ThrashedParanoid, Tsemii, Tymek, Urhixidur, Urmas, Volunteer Marek, WLKP, Wik, William Allen Simpson, XpertPL, Јованвб, 24 anonymous edits

Jogaila *Source*: http://en.wikipedia.org/w/index.php?title=Jogaila *Contributors*: 1812ahill, Abune, Academic Challenger, Adam78, AdjustShift, AjaxSmack, Ajh1492, Alekjds, Alexcoldcasefan, Alma Pater, Alphaboi867, Altenmann, Angusmclellan, Anthony, Appleseed, ArielGold, Asalrifai, Associated Greenwich, Attilios, AuburnPilot, Ausir, Avraham, Balcer, Beaumont, Bender235, Bigmac31, Biruitorul, Bloomfield, Bonás, Brandmeister (old), Brandon97, Brighterorange, Caius2ga, Cautious, Centrx, Choihei, Chris the speller, Chromaticity, CloudNine, CommonsDelinker, CrniBombarder!!!, Curps, Cyde, Cyon, Czalex, D6, Dachshund, Danny, DarioTW, David Parker, Deacon of Pndapetzim, DeirYassin, DerHexer, Dimadick, Dr. Dan, DrKiernan, Dreamafter, Dustpelt96, Dysmorodrepanis, Eclecticology, Elendil's Heir, Elonka, EmanWilm, Emax, Epbr123, Estlandia, Ev, Exxess, Fang Aili, FayssalF, FeanorStar7, Finlandais, Flying Saucer, Fon, Francis Schonken, Fsotrain09, GTBacchus, Gaius Cornelius, Georgette2, Ghirlandajo, GiW, Graham87, Grimfist, Gustav Korwin-Szwedowski, Halibutt, Interswd, Intgr, Iridescent, Iulius, J04n, Jaki kisur, Jki321, John of Reading, Juraune, Kaldari, Karl-Henner, Kazkaskazkasako, Kjkolb, KnowledgeOfSelf, Knutux, KonradWallenrod, Kozuch, Kpalion, Kpjas, Kummi, Kyriakos, Layla93, Leo1410, Lightmouse, Ling.Nut, Logologist, Lokyz, Lord Emsworth, LordXander1415, Lorielle, Lysy, M.K, Maed, Maglocunus, Marrtel, Mathiasrex, Matthead, Mattis, Mic, Michael Devore, Moxfyre, Moxica36, Ms2ger, Naive cynic, Naravorgaara, Nicke Lilltroll, Nightstallion, Nihil novi, Nino.shoshia, NorCalHistory, Odder, Olessi, Orionus, Outriggr, Pearle, Peter Isotalo, Pibwl, Pigman, Piotrus, Pixelface, PolishPoliticians, Przepla, Qp10qp, Quintote, R'n'B, Radomil, RandomP, Raudys, Remigiu, Renata3, Rich Farmbrough, RickK, Risk34, Rjwilmsi, RogDel, Romuald Wróblewski, Rosameliamartinez, Ruszewski, Rydel, Sannse, SchreiberBike, Schwartz und Weiss, Sean a. rose, Shilkanni, Showers, Sirmylesnagopaleentheda, Sk'py Skwrrrl, Smerus, Someone else, Space Cadet, Spacini, Sqwe, Srnec, Surtsicna, SylwiaS, Szopen, Tasc, ThomasPusch, Thoobik, Tim Starling, Topbanana, Tsuba, Ttwaring, Uris, Valentinian, Vecrumba, Vervin, Vladislaus Rex, Volunteer Marek, Wakeman12, Watch37264, Weisbrod, Wendell, Wknight94, Yarovit, Yomangani, Zello, Zillyonzhells, Zoe, Δ, Александър, 175 anonymous edits

Poland *Source*: http://en.wikipedia.org/w/index.php?title=Poland *Contributors*: 130.94.122.xxx, 217.99.96.xxx, 2eet2eet, 334a, 3sen, 4 bity muzyki, 4darek, 9cds, A bit iffy, A-giau, A. (pl), A.Jalil, A2Kafir, A8UDI, ABCD, ARC Gritt, Aaron Brenneman, Abductive, Abeg92, Academic Challenger, Achaemenes II, Acroterion, Adam Rock, AdamJacobMuller, AdamantlyMike, Adambro, Adamcisme123, Adashiel, Adavis444, Adz, Ae-a, Afinati, Agnus, Agquarx, Aherunar, Ahoerstemeier, AirdishStraus, AirlineMan333, Aivazovsky, Ajh1492, Akamad, Akanemoto, Alawadhi3000, Albanman, Alcatraz014, AlefZet, Aleph-4, Aleph4, Alex.muller, Alex1100, Alexandria, Ali@gwc.org.uk, AliveFreeHappy, AllenZh, Allstarecho, Almankib, Altenmann, Amakuru, Amish 01, Ammon86, Amoruso, Ancheta Wis, Andre Engels, Andy M. Wang, Andy Marchbanks, Andy4789, Andycjp, Andypl, AngelOfSadness, Angela, Anger22, Angr, Annarapo, Annonnimus, Anonymous editor, Anonymous from the 21st century, Another disinterested reader, Antakaa, Antandrus, Antique Rose, Antman, Aotearoa, AppleJordan, Appleseed, Aquintero82, Architect JDF, Arcillaroja, Argentina 678, Ari x, ArielGold, Aries1975, Arjun01, Army1987, Armydepot, Art LaPella, Arthena, Arvis21, Arwel Parry, AshLOVESgnr, Asidemes, Astian, Ataltane, Atlant, Atomekk, Attilios, Atwardow, Audirs8, Auntof6, Ausir, Austrian, Avala, Avaya1, Avb, Avicennasis, AxG, Axegod12, Axt, AzaToth, Azalero, Azar66, AzseicsoK, B-2StealthEditor, BGManofID, BLT000, BRG, Babajobu, Backspace, Balcer, Bardhylius, Baristarim, Barry Kent, Barryob, Bart133, Barticus88, Bartsimpson, Baseball Watcher, Bastian17, Battlecry, Bayerischermann, Bazonka, Bazza 7, Bbatsell, Bbulzibar, Beagel, Beamathan, Beaumont, Beefyt, Beegees, Beetstra, Bejnar, Beland, Ben-Velvel, Bencherlite, Benqish, Benwildeboer, Berkunt, BernardZ, BernardaAlba, Betacommand, Bftravel, Biala Gwiazda, Bigboydog1, Bility, Bill3000, Bjung, Bkell, Black Falcon, Blair P. Houghton, Blood Red Sandman, BlueMars, Blueshade, Bmicomp, BoBo1488, Bob marley 118, Bobblewik, Bobbybrown07, Bobo192, Bobtbuilder, Bocianski, Bodhi141, Boeing720, Bogdan, Bolo1729, Boraczek, Borgarde, Borisblue, Bornkaren, BostonMA, Boud, Boydenmb, Bprsolt Qaoddz, Br77rino, Brabblebrex, Brandmeister (old), Brandon, Briaboru, Brian0918, Brion VIBBER, Brisvegas, BritishWatcher, Brudasek, Brunopioss, Brutannica, Buaidh, Bukubku, Bull Market, BurgererSF, Burschenschafter, Bushy moustache, Bwv1004, C filev, C.Fred, CHJL, CIreland, CLW, CT Cooper, CTF83!, CWY2190, Cactus.man, Caerwine, Caius2ga, Caknuck, CalJW, CalicoCatLover, Caltas, CambridgeBayWeather, Can't sleep, clown will eat me, CanadianPride, CanisRufus, Canterbury Tail, Cantus, Caponer, CaptainFugu, Carbonix, Carlwev, CarolGray, Casliber, Caspian, Cassini83, Catherine-Jayne, Cautious, Celinakat, Celtmist, Century0, CesarB, ChKa, Chabuk, Challisrussia, CharlesMartel, Charlietemps, CharlotteWebb, Chase me ladies, I'm the Cavalry, Cheesefondue, Cheeseloop, ChemTerm, Chensiyuan, Chevinki, Chris G, Chris cufc, Chris the speller, ChrisGriswold, Chrisch, Chrischrischrischris, Chrishmt0423, Chrislk02, Christian List, Christoph.mamok, Christopher Kraus, Chriswilkins2, Chromaticity, Chudzik321, Chumchum7, Chunky Rice, CieloEstrellado, Cimmerian praetor, Civil Engineer III, Closedmouth, CloutierFan02, Cocoaguy, Colchicum, Colonel Mustard, CommonsDelinker, Conscious, Consequencefree, Constantine Gorov, Constantzeanu, Conversion script, Coolszach2, Cormic, Corpx, Corvettegrandsport, Count Iblis, Coveritus, Covervisit, Cptnono, Craftypants, Crazycomputers, Crystallina, Cs-wolves, Csiew88, Cubus1984, Cunningham, Curps, Cusio, Cybergoth, Cybershore, Czas Na Zywiec, Czrisher, Czyrko, D, D6, DBaba, DJ Clayworth, DJPeugeot, DMJohnston, DO'Neil, DRoninLA, DSuser, DVD R W, DX-Kevo, Daddy1234, Daffy123, Damian Radu, Damiang1997, Danaeb, Daniel De Mol, Daniel Quinlan, Daniil naumoff, Danim, Danlock2, Danny, Dannycas, Danziger100, Dariusz Szwed, Darklord517, Darth Kalwejt, Darwinek, Dave.Dunford, Davenbelle, Davewild, David Kernow, David Liuzzo, David Parker, David Richtman, Davidkazuhiro, Dawidbernard, Dbachmann, Dbenbenn, DeadEyeArrow, Deanna Bardenfleth, Deathmetal Greg, Debresser, Deflective, Dekisugi, Delldot, Demiurge, Denkealsobin, Denniss, Der Golem, DerHexer, Derek Ross, DerekJeter, Deschain97, Desiphral, Deuar, Digitalme, Digwuren, Dillard421, Dimkoa, Dina, Diogenes, Director958, Discospinster, Distal24, DivineIntervention, Dj3500, Dmanning, Dockingman, Docta247, Docu, Dodo von den Bergen, Dohaeng, Don Alessandro, Doopdoop, Doprendek, Dougy109, Downyourpants, DrSlony, Dragoman1, Draktorn, Dreadstar, Dru of Id, Drwhorules, DubaiTerminator, Dudeimoncoke, Dust Filter, DuttyYo, Duxwing, Dyamantese, Dycedarg, Dysepsion, Dysprosia, E Pluribus Anthony, EMT1871, ERcheck, ESkog, EagerToddler39, EarthPerson, Eclecticology, Edderso, EddyVadim, Editor at Large, Edivorce, Edward saint-ivan, Edward321, Ehheh, Ejoty, Ejército Rojo 1950, Ek8, Ekuprusevicius, El C, El-042, Eldin, Electionworld, ElectricValkyrie, Elementjunky, Elipongo, Elliskev, Elmondo21st, Elonka, Emax, Emigrant85, Emil Kastberg, Emote, Enchanter, EncycloPetey, Encyclopedist, Enforcer Of Lulz, Entryshare, Enviroboy, Enzo Aquarius, Eob, Ericamick, Erik1980, Esperant, Espoo, Etams, Eu.stefan, EugeneZelenko, Eumolpo, EuroLine, Ev, Evercat, Everyking, Everytime, Evlekis, FF2010, Fabhcún, Fabian Hassler, Farlieghhungerford, Farosdaughter, Farquaadhnchmn, Fasach Nua, FashionNugget, FayssalF, Fbeauregard, Fbifriday, Femto, Filemon, Filipdr, Filpaul, Filthdafugout, Filyo79, FinnishDriver, FireGaamer1122, Fireaxe888, Firstorm, FlagUploader, Flamarande, Flatterworld, Flockmeal, Fnagaton, Fordan, Fordie123, Foregone conclusion, Formeruser-81, Formulax, Foundert, Fox Mccloud, Fram, Francvs, Freak in the bunnysuit, Freakyfridge, Frees, FreplySpang, Freyr, Frietjes, Frstr, Frymaster, FunPika, Funandtrvl, Funnyhat, Future Perfect at Sunrise, Fvincent, G. Campbell, Gabbe, Gadfium, Gagabrain, Gaius Cornelius, Galati, Galileo01, Gam0001, Garion96, Garraisgreat, Garzo, Gaz, Gazilion, Gazza1133, Gdo01, Geeks123, Geeoharee, Gene Nygaard, Geneb1955, Generalboss3, Geniu, Geopolio, George The Dragon, Georgeslegloupier, Gggh, Gherald, Ghost321, GhostPirate, GhostWhoVotes, Gibble, Gilliam, Gimmetrow, GinaDana, Giraffedata, Giving is good, as long as you're getting, Gnesener1900, Gobonobo, Godfinger, Gogo Dodo, GoingBatty, Golbez, Golem Unity, Goltz, Good Olfactory, Goodlittleangel, GorillazFanAdam, GraemeL, Graham87, Gran2, Grayshi, Green Giant, GregorB, Grendelkhan, Greyhood, Ground Zero, Grsd, Grubber, Gryffindor, Grzechooo, Gscshoyru, Gsklee, Gtg204y, Guinness man, Gurch, Gustavo Szwedowski de Korwin, Gwernol, Gznornenplatz, Gzornenplatz, H, H3xx, H8erade, HIDECCHI001, HJ Mitchell, Hadal, Hagerman, Halibutt, Happyme22, Hardman1, Harmattan, Hatch68, Hayden120, Hdkhdk, Hdt83, Hebereke, Heimstern, Hello32020, Henry Flower, Hephaestos, Hillel, HistoriaPolska, Historian932, HistoricalAce, Hmains, Holod, Homunculus 2, Hoschimobi, Hottentot, Hu12, Husond, Hut 8.5, Ian Pitchford, Ianhowlett, Iapain wiki, Icehawk1223, Icollins68, Icseaturtles, Idaltu, Ief, Ignacy nowopolski, Ignatznice, Ikiroid, Iksisy, Illythr, Imagod777, Imlepid, InShaneee, Indisciplined, Indon, Infogengachi, Informationguy, Insanely Beautiful, InsanityJ, Intangible2.0, Inter, Interfector, Internet Ignoramus, Iohannes Animosus, Irek Biernat, Irishguy, Istros, Istvan, IstvanWolf, Itai, Italiano111, Ivan89, J.delanoy, J04n, JLaTondre, JRWalko, Jacek Kendysz, Jacoplane, Jacurek, Jadog9495, Jagoperson, Jahanas, JamesBWatson, Jameswilson, JanaMitz, Jankaspar, Janmariamazurczak, Janneman, Janus Shadowsong, JanusTehPheo, Japsai, Jaranda, Jarble, Jared Preston, Jaro7788, Jauerback, Jcaragonv, Jcc7386, Jcrook1987, JdeJ, Jecar, Jeff G., Jeff8765, Jeffhoy, Jeffmaylortx, Jensboot, Jerem43, Jh51681, Jhendin, Jiang, Jihg, Jiminy Krikkitt, Jimmyvanthach, Jirka.h23, Jizg, Jjbaruel, Jkelly, Jklin, Jkw0010, Jlujan69, Jni, JoeBlogsDord, Joffeloff, John, John K, John Price, John Smith M.D., Ph.D, John Vandenberg, John of Reading, Johnkenyon, Joho104, Jon513, Jonas Taylor, Jor, Jorunn, Joseph Solis in Australia, Josephschutz, Josiah Rowe, Jossi, Joy, Jpeob, Jpgordon, Jprg1966, Jrooksjr, Jseidel, Jsport, Jtbelliott, Jukes85, Julekmen, Julo, Jumbuck, Jura2102, Juraj cisarik, Jurand, Jusjih, Jwissick, Jxraynor, KConWiki, KJohansson, KOP, Kaihsu, Kajzderski, Kamil m, Kantokano, KapilTagore, Karasek, Karenjc, Karl, Karol Langner, Kaszkawal, Katarzyna, Kbrooks, Keegan, Keelm, KeithB, Kelisi, Kelly Martin, Kemiv, Khazar, Khoikhoi, King Bigos, Kinneyboy90, Kintetsubuffalo, Kirk diamond, Klimbert, Kman665, Kmcdm, Know Your NME, KnowledgeOfSelf, Knowledgebycoop, Knut, Knverma, Kobold 86, Kochamanita, KokkaShinto, Kollie1, KoloszRodos, Konkursor, Konrii, Konstable, Kopcik, Kotasik, Kotjze, Kotniski, Koyaanis Qatsi, Kozuch, Kpalion, Kpjas, KrakatoaKatie, Krautsaredummies, Krecik001, Kresen, Krich, Krupo, Krystiandl, Krzysiu, Krzyzowiec, Ksenon, Kuba1000, Kubek15, Kubigula, Kubster, Kukini, Kungfuadam, Kuracpm1, Kuru, Kusma, Kwamikagami, Kwpolska, Kyokpae, L95slovenia, LA2, LAX, LMB, LOL, LOTkid, LUCPOL, Laflin22, Laforgue, Lambrook, Lampak, Langus-TxT, Latka, Laudaka, Lcamtuf, Le Fou, Leafyplant, Lear 21, Lebanonman19, Lectonar, LeeHunter, Lemonade100, Lenin and McCarthy, LeonWhite, Levineps, Leyo, Liam Plested, LibLord, LifeStar, Liftarn, Lightdarkness, Lightmouse, Lights, Likeminas, Ling.Nut, Listowy, Llewelyn MT, Llort, LlywelynII, Lockesdonkey, Logan,

Logologist, Lolster, Lomisss, LonelyMarble, Looxix, Lord Voldemort, Loren36, Lost tourist, LoveToWiki, Lovesthespooge, Lrschum, Lst27, Luapnampahc, Luca Italy, Lucasm, Ludvikus, LugPaj, Luna Santin, Lycurgus, Lynntoniolondon, Lysy, M.C., MAXXX-309, MD, MER-C, MJCdetroit, MJM74, MMich, MSGJ, MaLu, MackSalmon, MadGeographer, Madriversnowboarder, Maelnuneb, Magister Mathematicae, Mailer diablo, Majorly, Maju wiki, Makemi, Makgraf, MakmoudHassan, Malcolm Farmer, Malcomosx, Malick78, Malik Shabazz, Malo, Mamaberry11, Man vyi, Maor X, Marcin Suwalczan, Marcin.sobczyk, Marco polo, Marek69, MarekF, Mareklug, Mark, MarkGallagher, Markotek, Martarius, Marthainky, Martim33, Martinwilke1980, Marx01, Marzoregulus@hotmail.com, Marzoregulus@yahoo.co.ph, Master of Puppets, Mathiasrex, Matman from Lublin, Matt 989, Mattbr, Matthead, Mauricio Maluff, Mav, Maxamegalon2000, Maxumum12, MałyG, Mb nl, Mc6415, Mc6809e, McDogm, Mcshadypl, Mdann52, Meekywiki, Mega Prime, Meika, Melizabeth, Meow kitties, Merbabu, Mercury, Mercurywoodrose, Mesgul82, Meursault2004, Mgiganteus1, Mic, Micga, Michael Hardy, Michael Zimmermann, Michalws, Mickey gfss2007, Micov, Microtony, Middle Drawer, Mieciu K, Migatu, Mike Rosoft, Miked2144, Mikenorton, Mikewebkist, Mild Bill Hiccup, Militaryboy, Milla297, Mimihitam, Minna Sora no Shita, Minpis, Mirek12, Mirv, Misiek889, Missmarple, Mix321, Miyokan, Miztalkalot, Mjanzen, Mjefm, Mnmazur, Modulatum, Mogism, Molobo, Momo50, Moncrief, Monkeyman, MonoSabio, MoogleEXE, Morwen, Mouse-Touch, Mps, Mr. Chainsaw, MrFish, Mrt3366, Msafiri, Mschel, Mushroom, Mvjs, Mx3, Myanw, Mysekurity, Myszodorn, NClement, NHRHS2010, Naddy, Naive cynic, Nakon, NameThatWorks, Naohiro19, Naryathegreat, Nasz, Natalie Erin, Natl1, NawlinWiki, Nayvik, Necronaut, Neferkare, Neurorocker, Neverquick, New4325, NewEnglandYankee, NewbieDoo, Ngchen, Niccohel, Niceguyedc, NickW557, Nickfraser, Nicolehere, Nietecza, NigelR, Nightstallion, Nihil novi, Nikkimaria, Niko226, Nima1024, Nite-Sirk, Niteowlneils, Nivix, Njaelkies Lea, Nksd, Nochoje, Nol888, Noozgroop, NorbertArthur, Norm mit, NormanBro, Norum, Notheruser, Notthe600, Nouanoua, NuclearVacuum, NuclearWarfare, Nufy8, Number 57, Numbo3, Nummymuffin, Nv8200p, Nyat, Nyyankees2fn, Oberst, Obli, Oda Mari, Odie5533, Offa of Angel, Ohnder, Ohnoitsjamie, Oktan, Old Moonraker, Oldgregg13, Oleg Alexandrov, Olga Raskolnikova, Olivier, Oliwier1111, Oly Majoieka, Omenge, Omicronpersei8, Omnibus, Oneiros, Oneness01, Only, Ontheglobe, Opelio, Opole.pl, Optimist on the run, Orange5716, Orczar, Oreo Priest, Ottawa4ever, Oxydo, Oxymoron83, P-Chan, PANONIAN, PFHLai, PG-Rated, Pabix, Pacus, Paddyez, Paine Ellsworth, Palosirkka, Pamejudd, Paragon12321, Parhamr, Parrottsvilleelementary, Pat Payne, Patientone, Paul Benjamin Austin, PaulHanson, PaulinSaudi, Paulomazzeirj, PawRon, Pawel bielsko, Pawel z Niepolomic, PawelO, Pawello071, Paweł ze Szczecina, Pawloo, PaxEquilibrium, Pb30, Pedant17, Pel'ican Sh'it, Pensionero, Pernambuko, Peruvianllama, Peter Chastain, Peter Horn, Peter558, PeterisP, Pethr, Petri Krohn, Phaedrus86, Phenz, Phgao, Philip Trueman, Phillip J, Philosopher, PhnomPencil, Picus viridis, Piepiepiepie2, Pilotguy, Pink!Teen, Piotr p928s, Piotrus, Pirawik, Pitdog, Placebo123, Plk, PlushMerga, Pmj, Podruznik, Poeticbent, PolEater, Polacoloco777, Polako, Polaron, PolishMan33, PolishPatriot, Polishtuner9, Politologia, Polski orzel, Polylerus, Porqin, Postdlf, Poszwa, Powertranz, PresN, PrestonH, Programming Hamster, Prusak, Prvc, Przemyśl, Przepla, Pseinstein, Pseudo-Richard, Psy guy, Ptmcd, PuRe$kilz.pl, Pudeo, Pundit, Pwt898, QaviArtilles, Qazwsx13, Qertis, Qogir, Qtoktok, Quadell, Quarma, Quebec99, Quendus, Queson, Quintote, Quizimodo (usurped), Qxz, R Lowry, R9tgokunks, RG2, RGA BALLY, RK, RL0919, Racejr, Radomil, Radzinski, Raficoo, Rajeshm1980, Ran, RapaNui75, Rarelibra, Ravelair, Raven4x4x, Ravenmoon22, Razer rob, RazorICE, Razum2010, Rbarreira, Rd232, Realberserker, RedArmyClub, RedMC, RedWolf, RedWordSmith, ReddShadoe, Redward1, Regulus marzo4103, Regulusmarzo@hotmai.l.com, Rejedef, Rejs12345, Remedios44, Retiono Virginian, Retired username, Rettetast, RexNL, Reywas92, Rfl, Rhys, Rich Farmbrough, Richard Harvey, Richardprins, Rick Block, Rickyrab, Rigby27, Rightfully in First Place, Riwnodennyk, Rje, Rjwilmsi, Rklisowski, Rkt2312, Rmhermen, Rmo13, Rmt2m, RoadTrain, Roastytoast, Rob Aleksandrowicz, Rob.derosa, Robenel, Robert Skyhawk, Robert Weemeyer, Robert346, RobertG, Robertgreer, Robertpolska, Robsuper, Roleplayer, Romanm, Romuald Wróblewski, Ronline, Rory096, Roryok, Rowingrocksmysocks, Rreagan007, Rudi Maxer, Ruhrjung, Rumping, Ruok, Ruszewski, Ruud64, Ruy Lopez, Rwerp, RxS, Ryanaxp, Rydia, Ryoho, Ryulong, Ryuu, Rübezahl, SCZenz, SJP, SOPHIAN, SPKirsch, SSJeskimobob, SSZ, SZPANER, Sadi, Sajimem, Salmon L-uh, Salvio giuliano, SamEV, SamWolken, Samah10, Samantha555, Samohyl Jan, Samrolken, Sanbeg, Sango123, Sannse, Sardanaphalus, Satyadasa, Savidan, Sca, Scarletpoet, Scewing, School is fun, Schopenhauer, SchreiberBike, Schwartz und Weiss, Science4sail, Scipius, Scohoust, Scott14, Scoutersig, Scwlong, Sean D Martin, Seb az86556, Sebastian770, Sebek2d, Sebjarod, Secret Saturdays, Secretlondon, Seka'sBreathMints, Selfworm, Sengkang, SenorAnderson, Ser Amantio di Nicolao, Seric2, Serte, SexBloodandMetal, Sfahey, Sfan00 IMG, Sfu, Shalom Alechem, Shanes, Shearonink, Sheitan, Shield, Shoeofdeath, Shoy, Shredder2012, Shuffdog, Shugo255, Sidume, Siemowit, Sietse Snel, Signalhead, Signsolid, Silar, Sillyfolkboy, Simon Peter Hughes, SimonP, Simonxag, Sirius85, Sjwk, Skarl the Drummer, Skatemanman, Skater1123, Skeithkiller, Skier Dude, Skizzik, Skooiman, Skoranka, Skuczera, Skurwiel, Slab stone, Sladen, Slakr, Slawojarek, Slecki19, Sliwers, Smith2006, Smitz, Smokizzy, Smyru, SnappingTurtle, SndrAndrss, Snowfalcon cu, Snowmanradio, Soman, SomeHuman, Someguy1221, Soroush83, Sortior, Soumyasch, South Bay, SpNeo, Space Cadet, Spamtroll, Spicy99, Spitzl, SpookyMulder, Springeragh, SqueakBox, Squids and Chips, Srikeit, Ssolbergj, Stanlavisbad, Starr37, Staszek Lem, Steel, Stefantastic, Stephenb, Steven Weston, Stickee, StoneProphet, Stressbaby 89, Stroganoff, StuMcDee, Subtropical-man, Suna, Sunholm, Sunquanliangxiuhao, Supertask, Supertouch, Surlyduff50, Surowiak, Surtsicna, Suva, Svetovid, SwPawel, SwPawel2, Swedish fusilier, Swid, SwisterTwister, Sycic, Sylwia Ufnalska, Symane, Syvanen, Szater, Szczecin, Szopen, TBadger, TFMcQ, TFOWR, TUF-KAT, TaerkastUA, Tandrasz, Tanessi, Targeman, Tassedethe, Tatry, TaulPaul, Tavrian, Taw, Tawker, Taxman, Tbhotch, Tblack93, Tcc8, TeaDrinker, Team Poland, Teddycom, Tekken50, Template namespace initialisation script, Teodor von Zalenzer Halde, Terfili, Terra Novus, Terrencekeenan, Teysz Kamieński, Th' Duck, Th1rt3en, ThaGrind, Thatdog, Thayts, The Article Creator, The Catholic Knight, The Letter J, The Rambling Man, The sunder king, TheHYPO, TheKMan, TheMaineliner, Theanthrope, Thecoldmidwest, Thehelpfulone, Thelosmets08, Themanwithoutapast, Therequiembellishere, Theroen, Thesoccermage, Thingg, This user has left wikipedia, Thiseye, ThomasK, Thomaspca, Thomehr, Thorsten1, Thryduulf, Thw1309, Tide rolls, TimComm, Timrollpickering, Tiptoety, Titoxd, Tkynerd, Tnick12, Tobby72, Tomaso, Tomasz J Kotarba, Tomirlik, Tommaisey, Tommyjs, Tone, Totti76, TownDown, Tpbradbury, Travelbird, Treisijs, Trekway.com, Trialsanderrors, TrueHistoryPoland, Tryingtomakethingsright, Tsujigiri, Tuckerresearch, Turmerick, Tuspm, Tvieth01, Tweenk, TycoonTim, TygerDawn, Tymek1988, Typobox, Ulmanor, Ulric1313, Ultratomio, Umedard, Underlying lk, UnicornTapestry, Upi, User86654, Usergreatpower, UtherSRG, Uxejn, Valenciano, Valentinian, Valermos, Vampain, Van helsing, VanCleave, Vaquero100, Vardion, VegaDark, Venariul, VenomousConcept, Ventusa, Viewfinder, Viriditas, Visor, Vitund, Vkem, Volunteer Marek, Von Fiszman, VonShroom, Vortexrealm, Vorthax, Vrinan, Vsmith, WGee, WKS Śląsk Wrocław, WOSlinker, WWEDawn, Waldir, Walton One, Wangi, Wavelength, WegianWarrior, Wehberf, WellsSouth, Weorthe, Werdan7, Werideatdusk33, West Brom 4ever, Wgitihm, WhisperToMe, Whomp, Wichid, Wik, Wiki alf, Wikibofh, Wikiborg, Wikidexel2, Wikidudeman, Wikiperuvian, Wilberforcehumphries, Will Beback Auto, William Allen Simpson, William Avery, Wilsonyork, Wimt, Witkacy, Wizzard2k, Wknight94, Woertche.Leo, WojPob, Wojgniew, WolfgangPeters, Wolfrock, Woohookitty, Wtmitchell, Wudkanz, Wulfstan, Wuzzy, Ww2censor, Wyklety, Władysław Komorek, XParadigm777x, XSmartCutie217, Xathaec, Xezbeth, Xlkce, Xtrump, Xx236, YUL89YYZ, Yakudza, Yamaguchi, Yamaka122, Yarovit, Yasis, Yattum, Yatzhek, Ybgursey, YellowMonkey, Yonas29, Yonna, Yorkshirian, Youstill let me make accounts lol, Zabadinho, Zanimum, Zaphod Beeblebrox, Zbihniew, Zello, Zenit4ever, ZimZalaBim, Zinjixmaggir, Zippy, Zlerman, Zoney, Zsinj, Zundark, Zzuuzz, Żlepskż, Δ, Александър, ГДБОБ, 3003 anonymous edits

Widawa, Łódź Voivodeship *Source*: http://en.wikipedia.org/w/index.php?title=Widawa%2C_%C5%81%C3%B3d%C5%BA_Voivodeship *Contributors*: Kotniski, Martinmdavis

Łask *Source*: http://en.wikipedia.org/w/index.php?title=%C5%81ask *Contributors*: Adam78, Akumiszcza, Appleseed, Balcer, Barticus88, Caius2ga, Chris G, CommonsDelinker, CryptoDerk, Darwinek, Emax, Gene Nygaard, Ike9898, JHunterJ, Kappa, Lysy, Marek69, Martinmdavis, Michael Hardy, Nono64, Sheynhertz-Unbayg, Sicherlich, Sourcelat0r, Susvolans, Weisbrod, YUL89YYZ, 3 anonymous edits

Łask County *Source*: http://en.wikipedia.org/w/index.php?title=%C5%81ask_County *Contributors*: Balcer, Gene Nygaard, Kotniski

Gmina Widawa *Source*: http://en.wikipedia.org/w/index.php?title=Gmina_Widawa *Contributors*: Darwinek, Martinmdavis

Village *Source*: http://en.wikipedia.org/w/index.php?title=Village *Contributors*: 1223Sallybride, 21655, A.Hassen, AKeen, Aarohan Mettu, Abductive, Acsian88, AdamDeanHall, Adamhauner, Adityamadhav83, Airjatt, Aitias, Akrabbim, Al Ameer son, Alaexis, Ale jrb, AlefZet, Aleksandr Grigoryev, Alensha, Altenmann, Amaury, Amberrock, Aminullah, Andrew M. Vachin, Andrewpmk, Andycjp, Animum, Apalamarchuk, Apokrif, Arpingstone, Arthena, AussieLegend, Azreey, Bejinhan, Beland, Belirac, Berig, Big Bird, Bill37212, Biruitorul, Bjh21, Bkonrad, Bldonne, Bobrayner, BocoROTH, Bogdan, Bogdangiusca, Bomac, Bossmanham13, Bossmanham98, Branka France, BreakingFree40, Brian the Editor, Brian0918, Buistr, Butko, C.Fred, Caiaffa, CalJW, CalumMcA, Canley, Capricorn42, Cassowary, Catgut, Catslisten, Cecropia, Cenarium, Ceriy, Cgoodwin, Chasnor15, Chentchen, Chirag, Chris j wood, Christopher.e.dunn, Chzz, Ciudad jardin, Cntras, Colonies Chris, CommonsDelinker, Conleyb15, Conorbrady.ie, Corregere, Cosmic Latte, CrimsonBlue, Cwolfsheep, D6, DITWIN GRIM, Da-drumer, DailyWikiHelp, Daniel Case, Daniel J. Leivick, Daniel Quinlan, David Biddulph, David Kernow, Dbg1233, Ddstretch, DePiep, Deathawk, Deeptrivia, Den fjättrade ankan, Deqon, Deror avi, Desiphral, Dezaree, Dieter Simon, Dinesh.jangra1, Discospinster, Djr13, Doctor Whom, Donarreiskoffer, Doseiai2, Dottod, Dpkipling, Durgesh.Rai, Dysepsion, E ponti, E0steven, Egard89, Ehrenkater, Elassint, Eleassar, Ellywa, Elockid, Elroygoh02, Emyr93, Epipelagic, Erianna, Eu.stefan, EugeneZelenko, EurekaLott, Extra999, Ezhiki, Fang 23, Farosdaughter, Fat pig73, Fayenatic london, Flewis, Francis Tyers, Francs2000, Fred Bauder, Frosted14, Fujurcitook, Funandtrvl, FunkyFly, Futurebird, Gardar Rurak, Garhowell, Gegnome, Gfoley4, Gingermint, Ginkgo100, Giro720, Glenn, Globalphilosophy, GoOdCoNtEnT, Gogo Dodo, Gorai, Grafikm fr, GrahamSmithe, Gtswiki, Guanaco, Guss2, Gustavo Szwedowski de Korwin, Gwen-chan, Gz33, Hahnchen, Halsteadk, Harisu1988, Hayabusa future, Headtoadman, Hebrides, Helenmwarwick, HenryLi, HighSpeed-X, Hmains, Hohum, Hroðulf, Hu12, Hydrogen Iodide, IGeMiNix, IShadowed, IW.HG, Ianisbored, Indiaforever, Ineffable3000, Inwind, Isam, IstvanWolf, IvanKlinko, J.delanoy, JHunterJ, JPSheridan, Jahangard, James086, Janetleopold, Jared Preston, Jcuk, JdeJ, Jeremy Bolwell, Jerrch, Jim1138, Jno, Jongleur100, Julesd, Julia W, JuneGloom07, KR3ColinB, Kajasudhakarababu, Kamiekam, Keelanrush, Kelisi, Kelovy, Khalid Mahmood, Khanvez, Khazar2, Kikos, Kimberry352, Kironbd07, KirrVlad, Kjramesh, Kman543210, Kmsasidhar, Komalrajiana, Ktr101, Kurruption, Kwamikagami, Larucio, Laurinavicius, Ldonna, Leonig Mig, Levineps, Lew121, LiDaobing, Lightmouse, LilHelpa, Linaduliban, Little muddy funkster, Littlealien182, Llywrch, Lonewolf BC, Lord Geznikor, Lozleader, Luk, Lulu1127, Lumos3, Lynda Finn, MacTire02, Maglame, Maitch, Majorly, Malick78, Mallakottai, Mani1, Manly 2008, Marek69, Mariano12 1989, Mariwan zanko, Mark91, Markmark12, Materialscientist, Mayooranathan, Mdzafri, Meldor, Menacheryarun, Mfo7787, MickMacNee, Mike Rosoft, MisfitToys, MisterVodka, Modulatum, Moebiusuibeom-en, Moonriddengirl, Mrmuk, Ms. Hayden and Ms. S, Mtaylor848, Mvjs, Mwalcoff, Mxn, NE2, Naturenet, Naveen Sankar, NerdyScienceDude, Nethergreen55, NewEnglandYankee, Nickshanks, Nistra, Noclevername, Northdeigo, Nricardo, Nyttend, O1ive, Old Moonraker, Orangemike, OscarKosy, P.Marlow, Pankajkumar.9009, Parkwells, Patrick, PaulBannister, Pawyilee, Penrithguy, Peter James, PeterCScott, Pgan002, Phagopsych, Pharaoh of the Wizards, PhilKnight, Pinethicket, PleaseStand, Pokrajac, Ppeace, Pug1776, Puneet 348, Quaysys, R'n'B, RDF, Raeky, Randhirreddy, RandomAct, Rcingham, Realteacher, Reeses14, Reinyday, Rich Farmbrough, Rickyrab, Risker, Rjm at sleepers, Robdav69, Rocket71048576, Rwxrwxrwx, Saccerzd, Sadrettin, Saga City, Same7same7, SamuraiClinton, Samyo, Sansonic, Sardanaphalus, Schinhofen, Schzmo, ScottSteiner, Seb az86556, Seivadnadroj, Sem10efg, Sephiroth BCR, Sfu, Shakingjoker117, Shamsurabiu, Shantavira, Sheddybey, Shell Kinney, Shergo korali, Shoeofdeath, SilkTork, Sjakkalle, Sk jmp, Sketchmoose, Sljaxon, Sole Soul, Speciate, StephenDawson, Steven Zhang, Stevo D, StewartNetAddict, Stifle, Student BSMU, Subhrat, Suffusion of Yellow, Suprgye, Surya Prakash.S.A., Szambo, THATSBETTER, Taamu, Tabletop, TaintedMustard, Tangerinehistry, Tbhotch, Teinesavaii, TejasDiscipulus2, TenIslands, Tenth Plague, TerryAga, The Thing That Should Not Be, The Transhumanist, TheArguer, Theranos, This, that and the other, Thomasrflagel, Tide rolls, Tjanke, Tobby72, Tobias Conradi, Tpbradbury, Triona, Tsuchiya Hikaru, TurkChan, Twanner921, Twin Bird, Tytyty456, Ukabia, UnicornTapestry, Untifler, Utcursch, V2k, VMS Mosaic, Vegaswikian, Vhdwikiuser, Vikramdidawat, Vivek Rai, Vmenkov, WOSlinker, Walex03, Warofdreams, Wayland, Wayne Slam, Wendy1991, West Brom 4ever, Whytecypress, Widr, Wiki Wikardo, Wikijens, William Allen Simpson, William Avery, Winhunter, Wknight94, WojPob, Woohookitty,

WorldWide Update, Wrelwser43, Xiaoyu of Yuxi, Xyz or die, Yakudza, Yonkie, Yupik, Zacharie Grossen, Zloyvolsheb, Zoeyzazu, Zollerriia, Zotel, Александър, პაატა შ, , , , 471 anonymous edits

Image Sources, Licenses and Contributors

file:Poland location map.svg *Source*: http://en.wikipedia.org/w/index.php?title=File:Poland_location_map.svg *License*: unknown *Contributors*: User:NordNordWest

File:Red pog.svg *Source*: http://en.wikipedia.org/w/index.php?title=File:Red_pog.svg *License*: unknown *Contributors*: Anomie

File:Flag of Poland.svg *Source*: http://en.wikipedia.org/w/index.php?title=File:Flag_of_Poland.svg *License*: unknown *Contributors*: Anomie, Mifter

File:Collage of views of Lodz.PNG *Source*: http://en.wikipedia.org/w/index.php?title=File:Collage_of_views_of_Lodz.PNG *License*: unknown *Contributors*: User:KoloFortuny

File:POL Łódź flag.svg *Source*: http://en.wikipedia.org/w/index.php?title=File:POL_Łódź_flag.svg *License*: unknown *Contributors*: user:MesserWoland

File:POL Łódź COA.svg *Source*: http://en.wikipedia.org/w/index.php?title=File:POL_Łódź_COA.svg *License*: unknown *Contributors*: user:Poznaniak

File:Speaker Icon.svg *Source*: http://en.wikipedia.org/w/index.php?title=File:Speaker_Icon.svg *License*: unknown *Contributors*: Blast, G.Hagedorn, Mobius, Tehdog, 2 anonymous edits

File:Archikatedra Łódź v2.jpg *Source*: http://en.wikipedia.org/w/index.php?title=File:Archikatedra_Łódź_v2.jpg *License*: unknown *Contributors*: user:HuBar

File:Stacja Radegast pomnik1 Łódź.JPG *Source*: http://en.wikipedia.org/w/index.php?title=File:Stacja_Radegast_pomnik1_Łódź.JPG *License*: unknown *Contributors*: User:Zasina

File:Fontanna lodz.jpg *Source*: http://en.wikipedia.org/w/index.php?title=File:Fontanna_lodz.jpg *License*: unknown *Contributors*: User:Centymetr

File:Lodz UlPiotr sun.jpg *Source*: http://en.wikipedia.org/w/index.php?title=File:Lodz_UlPiotr_sun.jpg *License*: unknown *Contributors*: Original uploader was Wurzeller at en.wikipedia

File:Centrum Sienkiewicza Łódź.JPG *Source*: http://en.wikipedia.org/w/index.php?title=File:Centrum_Sienkiewicza_Łódź.JPG *License*: unknown *Contributors*: User:Zasina

File:Pomnik Rubisteina Lodz.jpg *Source*: http://en.wikipedia.org/w/index.php?title=File:Pomnik_Rubisteina_Lodz.jpg *License*: unknown *Contributors*: User:Polimerek

File:Rektorat Politechnika L.jpg *Source*: http://en.wikipedia.org/w/index.php?title=File:Rektorat_Politechnika_L.jpg *License*: unknown *Contributors*: User:Polimerek

File:Palac Oskara Kona Lodz.jpg *Source*: http://en.wikipedia.org/w/index.php?title=File:Palac_Oskara_Kona_Lodz.jpg *License*: unknown *Contributors*: User:HuBar

File:Arena-Lodz-mecz.jpg *Source*: http://en.wikipedia.org/w/index.php?title=File:Arena-Lodz-mecz.jpg *License*: unknown *Contributors*: User:Walther

File:Flag of Germany.svg *Source*: http://en.wikipedia.org/w/index.php?title=File:Flag_of_Germany.svg *License*: unknown *Contributors*: Anomie

File:Flag of Georgia.svg *Source*: http://en.wikipedia.org/w/index.php?title=File:Flag_of_Georgia.svg *License*: unknown *Contributors*: User:SKopp

File:Flag of Portugal.svg *Source*: http://en.wikipedia.org/w/index.php?title=File:Flag_of_Portugal.svg *License*: unknown *Contributors*: Columbano Bordalo Pinheiro (1910; generic design); Vítor Luís Rodrigues; António Martins-Tuválkin (2004; this specific vector set: see sources)

File:Flag of France.svg *Source*: http://en.wikipedia.org/w/index.php?title=File:Flag_of_France.svg *License*: unknown *Contributors*: Anomie

File:Flag of Finland.svg *Source*: http://en.wikipedia.org/w/index.php?title=File:Flag_of_Finland.svg *License*: unknown *Contributors*: User:SKopp

File:Flag of Lithuania.svg *Source*: http://en.wikipedia.org/w/index.php?title=File:Flag_of_Lithuania.svg *License*: unknown *Contributors*: User:SKopp

File:Flag of Mexico.svg *Source*: http://en.wikipedia.org/w/index.php?title=File:Flag_of_Mexico.svg *License*: unknown *Contributors*: User:AlexCovarrubias

File:Flag of Russia.svg *Source*: http://en.wikipedia.org/w/index.php?title=File:Flag_of_Russia.svg *License*: unknown *Contributors*: Anomie, Zscout370

File:Flag of Spain.svg *Source*: http://en.wikipedia.org/w/index.php?title=File:Flag_of_Spain.svg *License*: unknown *Contributors*: Anomie

File:Flag of Sweden.svg *Source*: http://en.wikipedia.org/w/index.php?title=File:Flag_of_Sweden.svg *License*: unknown *Contributors*: Anomie

File:Flag of Belarus.svg *Source*: http://en.wikipedia.org/w/index.php?title=File:Flag_of_Belarus.svg *License*: unknown *Contributors*: User:Zscout370

File:Flag of Ukraine.svg *Source*: http://en.wikipedia.org/w/index.php?title=File:Flag_of_Ukraine.svg *License*: unknown *Contributors*: User:Jon Harald Søby, User:Zscout370

File:Flag of Hungary.svg *Source*: http://en.wikipedia.org/w/index.php?title=File:Flag_of_Hungary.svg *License*: unknown *Contributors*: User:SKopp

File:Flag of Israel.svg *Source*: http://en.wikipedia.org/w/index.php?title=File:Flag_of_Israel.svg *License*: unknown *Contributors*: "The Provisional Council of State Proclamation of the Flag of the State of Israel" of 25 Tishrei 5709 (28 October 1948) provides the official specification for the design of the Israeli flag. The color of the Magen David and the stripes of the Israeli flag is not precisely specified by the above legislation. The color depicted in the current version of the image is typical of flags used in Israel today, although individual flags can and do vary. The flag legislation officially specifies dimensions of 220 cm × 160 cm. However, the sizes of actual flags vary (although the aspect ratio is usually retained).

File:Flag of Argentina.svg *Source*: http://en.wikipedia.org/w/index.php?title=File:Flag_of_Argentina.svg *License*: unknown *Contributors*: User:Dbenbenn

File:Flag of the People's Republic of China.svg *Source*: http://en.wikipedia.org/w/index.php?title=File:Flag_of_the_People's_Republic_of_China.svg *License*: unknown *Contributors*: User:Denelson83, User:SKopp, User:Shizhao, User:Zscout370

File:Flag of the United States.svg *Source*: http://en.wikipedia.org/w/index.php?title=File:Flag_of_the_United_States.svg *License*: unknown *Contributors*: Anomie

File:Palac Juliusza Heinzla Lodz.JPG *Source*: http://en.wikipedia.org/w/index.php?title=File:Palac_Juliusza_Heinzla_Lodz.JPG *License*: unknown *Contributors*: User:HuBar

File:Rynek S Miasta Lodz.jpg *Source*: http://en.wikipedia.org/w/index.php?title=File:Rynek_S_Miasta_Lodz.jpg *License*: unknown *Contributors*: User:Polimerek

File:Schweikert_Palace.jpg *Source*: http://en.wikipedia.org/w/index.php?title=File:Schweikert_Palace.jpg *License*: unknown *Contributors*: User:Polimerek

File:Lodz manufaktura 01.jpg *Source*: http://en.wikipedia.org/w/index.php?title=File:Lodz_manufaktura_01.jpg *License*: unknown *Contributors*: Artur Andrzej, Darekm135, HuBar, 3 anonymous edits

File:Pałac Karola Poznańskiego Łódź.JPG *Source*: http://en.wikipedia.org/w/index.php?title=File:Pałac_Karola_Poznańskiego_Łódź.JPG *License*: unknown *Contributors*: User:Zasina

File:Biala Fabryka2.jpg *Source*: http://en.wikipedia.org/w/index.php?title=File:Biala_Fabryka2.jpg *License*: unknown *Contributors*: User:Polimerek

File:WielkaSynagoga3 Lodz.jpg *Source*: http://en.wikipedia.org/w/index.php?title=File:WielkaSynagoga3_Lodz.jpg *License*: unknown *Contributors*: HuBar, Olve Utne, Sicherlich, Slav, 1 anonymous edits

File:1853 Łódź.jpg *Source*: http://en.wikipedia.org/w/index.php?title=File:1853_Łódź.jpg *License*: unknown *Contributors*: User:Sicherlich

File:Lodz1577.jpg *Source*: http://en.wikipedia.org/w/index.php?title=File:Lodz1577.jpg *License*: unknown *Contributors*: User:Sicherlich

File:Łódź - Pałac Izraela Poznańskiego.jpg *Source*: http://en.wikipedia.org/w/index.php?title=File:Łódź_-_Pałac_Izraela_Poznańskiego.jpg *License*: unknown *Contributors*: User:Lestath

File:Cerkiew Aleksandra Newskiego w Łodzi2.jpg *Source*: http://en.wikipedia.org/w/index.php?title=File:Cerkiew_Aleksandra_Newskiego_w_Łodzi2.jpg *License*: unknown *Contributors*: User:Arewicz

File:Scheibler-Mausoleum Lodz 2006.jpg *Source*: http://en.wikipedia.org/w/index.php?title=File:Scheibler-Mausoleum_Lodz_2006.jpg *License*: unknown *Contributors*: User:A.fiedler

File:2005-09 Łódź Narutowicza 20-22 (2).jpg *Source*: http://en.wikipedia.org/w/index.php?title=File:2005-09_Łódź_Narutowicza_20-22_(2).jpg *License*: unknown *Contributors*: User:Sicherlich

File:Park Poniatowskiego 1 Łódź.JPG *Source*: http://en.wikipedia.org/w/index.php?title=File:Park_Poniatowskiego_1_Łódź.JPG *License*: unknown *Contributors*: User:Zasina

File:LodzDocelowyUkladDrog.svg *Source*: http://en.wikipedia.org/w/index.php?title=File:LodzDocelowyUkladDrog.svg *License*: unknown *Contributors*: User:Sliwers, user:MesserWoland

File:Collage of views of Warsaw 1.png *Source*: http://en.wikipedia.org/w/index.php?title=File:Collage_of_views_of_Warsaw_1.png *License*: unknown *Contributors*: User:BurgererSF, User:Martimar

File:Flag of Warsaw.svg *Source*: http://en.wikipedia.org/w/index.php?title=File:Flag_of_Warsaw.svg *License*: unknown *Contributors*: User:Dbenbenn, User:Halibutt, User:Rfl

File:Warsaw emblem.png *Source*: http://en.wikipedia.org/w/index.php?title=File:Warsaw_emblem.png *License*: unknown *Contributors*: AtonX, Chepry, Gustavo Szwedowski de Korwin, Halibutt, Kilom691, Kuba G, L.m.k, Maikking, Man vyi, Mir, Poznaniak, Raf24, Sicherlich, Skim, Twthmoses, 1 anonymous edits

File:Kosciol NMP Warszawa Nowe Miasto.jpg *Source*: http://en.wikipedia.org/w/index.php?title=File:Kosciol_NMP_Warszawa_Nowe_Miasto.jpg *License*: unknown *Contributors*: User:Asia Tomczak

File:Dahlbergh Bitwa Warszawa 1656.jpg *Source*: http://en.wikipedia.org/w/index.php?title=File:Dahlbergh_Bitwa_Warszawa_1656.jpg *License*: unknown *Contributors*: BurgererSF, Esquilo, Hiuppo, Piotrus

File:German airship bombing Warsaw.JPG *Source*: http://en.wikipedia.org/w/index.php?title=File:German_airship_bombing_Warsaw.JPG *License*: unknown *Contributors*: Artur Andrzej, Auntof6, BurgererSF, Herbythyme, PeterWD, 2 anonymous edits

File:Destroyed Warsaw, capital of Poland, January 1945.jpg *Source*: http://en.wikipedia.org/w/index.php?title=File:Destroyed_Warsaw,_capital_of_Poland,_January_1945.jpg *License*: unknown *Contributors*: M. Swierczynski

File:Msza św na Pl Zwycięstwa.jpg *Source*: http://en.wikipedia.org/w/index.php?title=File:Msza_św_na_Pl_Zwycięstwa.jpg *License*: unknown *Contributors*: Daczor, Martimar

File:Varsaw SPOT 1066.jpg *Source*: http://en.wikipedia.org/w/index.php?title=File:Varsaw_SPOT_1066.jpg *License*: unknown *Contributors*: User:Spot Image

File:Warszawa outline with districts v2.svg *Source*: http://en.wikipedia.org/w/index.php?title=File:Warszawa_outline_with_districts_v2.svg *License*: unknown *Contributors*: User:Mfloryan

File:Warszawa-skyline-pole mokotowskie.jpg *Source*: http://en.wikipedia.org/w/index.php?title=File:Warszawa-skyline-pole_mokotowskie.jpg *License*: unknown *Contributors*: See and current image history.

File:Poland Warsaw Łazienki Palace.jpg *Source*: http://en.wikipedia.org/w/index.php?title=File:Poland_Warsaw_Łazienki_Palace.jpg *License*: unknown *Contributors*: User:Wojsyl

File:6 Warszawa 149.jpg *Source*: http://en.wikipedia.org/w/index.php?title=File:6_Warszawa_149.jpg *License*: unknown *Contributors*: Jarekt, Sfu, Shalom Alechem

File:Socreal decoration in Warsaw.JPG *Source*: http://en.wikipedia.org/w/index.php?title=File:Socreal_decoration_in_Warsaw.JPG *License*: unknown *Contributors*: Ala z, Halibutt, Panek, Piotrus, Sfu, Starscream, 4 anonymous edits

File:5 Warszawa 071.jpg *Source*: http://en.wikipedia.org/w/index.php?title=File:5_Warszawa_071.jpg *License*: unknown *Contributors*: Dysmorodrepanis, Jarekt, Panek, Sfu, Shalom Alechem, ToSter

File:Nowa Pomarańczarnia Łazienki Królewskie Warszawa 05.JPG *Source*: http://en.wikipedia.org/w/index.php?title=File:Nowa_Pomarańczarnia_Łazienki_Królewskie_Warszawa_05.JPG *License*: unknown *Contributors*: User:Boston9

File:Poland Warsaw boundaries 1939 and 2005.svg *Source*: http://en.wikipedia.org/w/index.php?title=File:Poland_Warsaw_boundaries_1939_and_2005.svg *License*: unknown *Contributors*: Marcin Floryan

File:7 Warszawa 163.jpg *Source*: http://en.wikipedia.org/w/index.php?title=File:7_Warszawa_163.jpg *License*: unknown *Contributors*: Beao, BurgererSF, Erik Baas, ToSter, Wedlowski

File:Sejm RP.jpg *Source*: http://en.wikipedia.org/w/index.php?title=File:Sejm_RP.jpg *License*: unknown *Contributors*: User:Kpalion

File:Warsaw-centrum.jpg *Source*: http://en.wikipedia.org/w/index.php?title=File:Warsaw-centrum.jpg *License*: unknown *Contributors*: Bpa

File:Warsaw Metro Plac Wilsona 2.jpg *Source*: http://en.wikipedia.org/w/index.php?title=File:Warsaw_Metro_Plac_Wilsona_2.jpg *License*: unknown *Contributors*: A.Savin, Admirał Bang, Anniolek, Apalsola, Dodo, Gnesener1900, Hiuppo, Infrogmation, Jcornelius, Nilfanion, Pko, Sfu, 2 anonymous edits

File:Widok1 cob.jpg *Source*: http://en.wikipedia.org/w/index.php?title=File:Widok1_cob.jpg *License*: unknown *Contributors*: Lech i Wojciech Szymborscy

File:Tress-Jazz-in-Tygmont-AB.jpg *Source*: http://en.wikipedia.org/w/index.php?title=File:Tress-Jazz-in-Tygmont-AB.jpg *License*: unknown *Contributors*: Chepry, Dr. Shaggeman, Eusebius, Infrogmation, 1 anonymous edits

File:Warsaw Multimedia Fountain Park 1.JPG *Source*: http://en.wikipedia.org/w/index.php?title=File:Warsaw_Multimedia_Fountain_Park_1.JPG *License*: unknown *Contributors*: User:Boston9

File:Moderne wolkenkrabber Warschau 0865.PNG *Source*: http://en.wikipedia.org/w/index.php?title=File:Moderne_wolkenkrabber_Warschau_0865.PNG *License*: unknown *Contributors*: User:VanNouten

File:Warszawa Teatr Wielki rzut prostoliniowy.jpg *Source*: http://en.wikipedia.org/w/index.php?title=File:Warszawa_Teatr_Wielki_rzut_prostoliniowy.jpg *License*: unknown *Contributors*: User:Sfu

File:074 Palace of Culture.jpg *Source*: http://en.wikipedia.org/w/index.php?title=File:074_Palace_of_Culture.jpg *License*: unknown *Contributors*: Frank Spakowski

File:Warszawa-MPW.jpg *Source*: http://en.wikipedia.org/w/index.php?title=File:Warszawa-MPW.jpg *License*: unknown *Contributors*: user:sfu

File:Ostrogski Palace Chopin Museum June 2010 d.jpg *Source*: http://en.wikipedia.org/w/index.php?title=File:Ostrogski_Palace_Chopin_Museum_June_2010_d.jpg *License*: unknown *Contributors*: Szczebrzeszynski

File:Moderne wolkenkrabber Warschau 0875.PNG *Source*: http://en.wikipedia.org/w/index.php?title=File:Moderne_wolkenkrabber_Warschau_0875.PNG *License*: unknown *Contributors*: User:VanNouten

File:6 Warszawa 159.jpg *Source*: http://en.wikipedia.org/w/index.php?title=File:6_Warszawa_159.jpg *License*: unknown *Contributors*: Hiuppo, Jarekt, Sfu, Shalom Alechem

File:6 Warszawa 407.jpg *Source*: http://en.wikipedia.org/w/index.php?title=File:6_Warszawa_407.jpg *License*: unknown *Contributors*: Marek i Ewa Wojciechowscy

File:Warsaw skyline.jpg *Source*: http://en.wikipedia.org/w/index.php?title=File:Warsaw_skyline.jpg *License*: unknown *Contributors*: User:BurgererSF

File:Moderne wolkenkrabber Warschau 0883.PNG *Source*: http://en.wikipedia.org/w/index.php?title=File:Moderne_wolkenkrabber_Warschau_0883.PNG *License*: unknown *Contributors*: User:VanNouten

File:169 2 a.jpg *Source*: http://en.wikipedia.org/w/index.php?title=File:169_2_a.jpg *License*: unknown *Contributors*: Martimar, Shalom Alechem, 2 anonymous edits

File:Warsaw - Royal Castle Square.jpg *Source*: http://en.wikipedia.org/w/index.php?title=File:Warsaw_-_Royal_Castle_Square.jpg *License*: unknown *Contributors*: Shalom Alechem

File:Warsaw Sirene 1659.PNG *Source*: http://en.wikipedia.org/w/index.php?title=File:Warsaw_Sirene_1659.PNG *License*: unknown *Contributors*: Kocio, Martimar, Mathiasrex, Picus viridis, Skim, 1 anonymous edits

File:Syrenka.jpg *Source*: http://en.wikipedia.org/w/index.php?title=File:Syrenka.jpg *License*: unknown *Contributors*: User:ABX

File:Mariecurie.jpg *Source*: http://en.wikipedia.org/w/index.php?title=File:Mariecurie.jpg *License*: unknown *Contributors*: Deror avi, Editor at Large, Jarekt, OsamaK, Savh, Tryphon, UpstateNYer, Wouterhagens, 10 anonymous edits

File:Flag of Kazakhstan.svg *Source*: http://en.wikipedia.org/w/index.php?title=File:Flag_of_Kazakhstan.svg *License*: unknown *Contributors*: -xfi-

File:Flag of Thailand.svg *Source*: http://en.wikipedia.org/w/index.php?title=File:Flag_of_Thailand.svg *License*: unknown *Contributors*: User:Zscout370

File:Flag of the United Kingdom.svg *Source*: http://en.wikipedia.org/w/index.php?title=File:Flag_of_the_United_Kingdom.svg *License*: unknown *Contributors*: Anomie, Good Olfactory, Mifter

File:Flag of the Netherlands.svg *Source*: http://en.wikipedia.org/w/index.php?title=File:Flag_of_the_Netherlands.svg *License*: unknown *Contributors*: User:Zscout370

File:Flag of Japan.svg *Source*: http://en.wikipedia.org/w/index.php?title=File:Flag_of_Japan.svg *License*: unknown *Contributors*: Anomie

File:Flag of Vietnam.svg *Source*: http://en.wikipedia.org/w/index.php?title=File:Flag_of_Vietnam.svg *License*: unknown *Contributors*: user:Lưu Ly

File:Flag of Turkey.svg *Source*: http://en.wikipedia.org/w/index.php?title=File:Flag_of_Turkey.svg *License*: unknown *Contributors*: User:Dbenbenn

File:Flag of Norway.svg *Source*: http://en.wikipedia.org/w/index.php?title=File:Flag_of_Norway.svg *License*: unknown *Contributors*: User:Dbenbenn

File:Flag of Brazil.svg *Source*: http://en.wikipedia.org/w/index.php?title=File:Flag_of_Brazil.svg *License*: unknown *Contributors*: Anomie

File:Flag of Latvia.svg *Source*: http://en.wikipedia.org/w/index.php?title=File:Flag_of_Latvia.svg *License*: unknown *Contributors*: User:SKopp

File:Flag of South Korea.svg *Source*: http://en.wikipedia.org/w/index.php?title=File:Flag_of_South_Korea.svg *License*: unknown *Contributors*: Various

File:Flag of the Republic of China.svg *Source*: http://en.wikipedia.org/w/index.php?title=File:Flag_of_the_Republic_of_China.svg *License*: unknown *Contributors*: 555, Abner1069, Bestalex, Bigmorr, Denelson83, Ed veg, Gzdavidwong, Herbythyme, Isletakee, Kakoui, Kallerna, Kibinsky, Mattes, Mizunoryu, Neq00, Nickpo, Nightstallion, Odder, Pymouss, R.O.C, Reisio, Reuvenk, Rkt2312, Rocket000, Runningfridgesrule, Samwingkit, Sasha Krotov, Shizhao, Tabasco, Vzb83, Wrightbus, ZooFari, Zscout370, 75 anonymous edits

File:Flag of Canada.svg *Source*: http://en.wikipedia.org/w/index.php?title=File:Flag_of_Canada.svg *License*: unknown *Contributors*: Anomie

File:Flag of Austria.svg *Source*: http://en.wikipedia.org/w/index.php?title=File:Flag_of_Austria.svg *License*: unknown *Contributors*: User:SKopp

Image:Elementy herbu.jpg *Source*: http://en.wikipedia.org/w/index.php?title=File:Elementy_herbu.jpg *License*: unknown *Contributors*: User:Qyd, User:Steifer

File:Hyghalmen Roll Late 1400s.jpg *Source*: http://en.wikipedia.org/w/index.php?title=File:Hyghalmen_Roll_Late_1400s.jpg *License*: unknown *Contributors*: AndreasPraefcke, Chatsam, Evadb, G.dallorto, Knorrepoes, LeoDavid, Massimop, Para, Skim, Technokrat, 1 anonymous edits

Image:University of Plymouth Coat of Arms.png *Source*: http://en.wikipedia.org/w/index.php?title=File:University_of_Plymouth_Coat_of_Arms.png *License*: unknown *Contributors*: Garter, Clarenceux and Norroy and Ulster King of Arms.

File:Vaasa.vaakuna.svg *Source*: http://en.wikipedia.org/w/index.php?title=File:Vaasa.vaakuna.svg *License*: unknown *Contributors*: User:Care

File:US-GreatSeal-Obverse.svg *Source*: http://en.wikipedia.org/w/index.php?title=File:US-GreatSeal-Obverse.svg *License*: unknown *Contributors*: U.S. Government

File:John paul 2 coa.svg *Source*: http://en.wikipedia.org/w/index.php?title=File:John_paul_2_coa.svg *License*: unknown *Contributors*: User:mAgul

Image:Coats of arms of the Kingdom of Egypt and Sudan.png *Source*: http://en.wikipedia.org/w/index.php?title=File:Coats_of_arms_of_the_Kingdom_of_Egypt_and_Sudan.png *License*: unknown *Contributors*: User:Beao, User:CrimeCentral, User:SreeBot

Image:Coat of arms of United Arab Republic (Syria 1958-61, Egypt 1958-1971).svg *Source*: http://en.wikipedia.org/w/index.php?title=File:Coat_of_arms_of_United_Arab_Republic_(Syria_1958-61,_Egypt_1958-1971).svg *License*: unknown *Contributors*: User:AnonMoos, User:Kafka1, User:Omar86, User:R-41, User:Tonyjeff

File:PD-icon.svg *Source*: http://en.wikipedia.org/w/index.php?title=File:PD-icon.svg *License*: unknown *Contributors*: Alex.muller, Anomie, Anonymous Dissident, CBM, MBisanz, PBS, Quadell, Rocket000, Strangerer, Timotheus Canens, 1 anonymous edits

File:Arms of Castile and Leon.svg *Source*: http://en.wikipedia.org/w/index.php?title=File:Arms_of_Castile_and_Leon.svg *License*: unknown *Contributors*: User:Heralder

File:Arms of Elizabeth Bowes-Lyon.svg *Source*: http://en.wikipedia.org/w/index.php?title=File:Arms_of_Elizabeth_Bowes-Lyon.svg *License*: unknown *Contributors*: User:Sodacan

File:Arms of Beatrice of York.svg *Source*: http://en.wikipedia.org/w/index.php?title=File:Arms_of_Beatrice_of_York.svg *License*: unknown *Contributors*: User:Sodacan

Image:Brasão de Berwickshire.png *Source*: http://en.wikipedia.org/w/index.php?title=File:Brasão_de_Berwickshire.png *License*: unknown *Contributors*: DenghiùComm, Jesielt, Massimop

Image:Elmbridge BC Crest of Arms.png *Source*: http://en.wikipedia.org/w/index.php?title=File:Elmbridge_BC_Crest_of_Arms.png *License*: unknown *Contributors*: Adam37

Image:Blason de la ville de Châteaurenard (13).svg *Source*: http://en.wikipedia.org/w/index.php?title=File:Blason_de_la_ville_de_Châteaurenard_(13).svg *License*: unknown *Contributors*: User:spax89

Image:Wappen Eberbach Baden.png *Source*: http://en.wikipedia.org/w/index.php?title=File:Wappen_Eberbach_Baden.png *License*: unknown *Contributors*: -

File:Hensbroek.svg *Source*: http://en.wikipedia.org/w/index.php?title=File:Hensbroek.svg *License*: unknown *Contributors*: User:Arch

File:FEC.png *Source*: http://en.wikipedia.org/w/index.php?title=File:FEC.png *License*: unknown *Contributors*: Brian Boru, Knorrepoes, Magul

Image:Falkenberg vapen.svg *Source*: http://en.wikipedia.org/w/index.php?title=File:Falkenberg_vapen.svg *License*: unknown *Contributors*: User:Marmelad

Image:Seinäjoki.vaakuna.svg *Source*: http://en.wikipedia.org/w/index.php?title=File:Seinäjoki.vaakuna.svg *License*: unknown *Contributors*: User:Hippophaë

File:Logo Lodzkie.jpg *Source*: http://en.wikipedia.org/w/index.php?title=File:Logo_Lodzkie.jpg *License*: unknown *Contributors*: Bushy moustache

File:POL województwo łódzkie flag.svg *Source*: http://en.wikipedia.org/w/index.php?title=File:POL_województwo_łódzkie_flag.svg *License*: unknown *Contributors*: user:Bastianow

File:POL_województwo_łódzkie_COA.svg *Source*: http://en.wikipedia.org/w/index.php?title=File:POL_województwo_łódzkie_COA.svg *License*: unknown *Contributors*: User:Bastianow

File:Lodzkie (EE,E NN,N).png *Source*: http://en.wikipedia.org/w/index.php?title=File:Lodzkie_(EE,E_NN,N).png *License*: unknown *Contributors*: user:Wulfstan, user:tsca

File:Wojewodzkie lodzkie adm.png *Source*: http://en.wikipedia.org/w/index.php?title=File:Wojewodzkie_lodzkie_adm.png *License*: unknown *Contributors*: MariuszR, Mir, Niki K, 1 anonymous edits

File:Lodz Voivodship 1938.png *Source*: http://en.wikipedia.org/w/index.php?title=File:Lodz_Voivodship_1938.png *License*: unknown *Contributors*: User:Halibutt

File:Lodz Voivodship 1975.png *Source*: http://en.wikipedia.org/w/index.php?title=File:Lodz_Voivodship_1975.png *License*: unknown *Contributors*: User:Halibutt

Image:Coat of Arms of Vladislav Jagaila as king of Poland.svg *Source*: http://en.wikipedia.org/w/index.php?title=File:Coat_of_Arms_of_Vladislav_Jagaila_as_king_of_Poland.svg *License*: unknown *Contributors*: User:Avalokitesvara, User:Bastianow, User:Orem, user:Sodacan, user:Steifer

File:Polska 1386 - 1434.png *Source*: http://en.wikipedia.org/w/index.php?title=File:Polska_1386_-_1434.png *License*: unknown *Contributors*: user:Poznaniak

Image:Grobowiec krolowej Jadwigi.jpg *Source*: http://en.wikipedia.org/w/index.php?title=File:Grobowiec_krolowej_Jadwigi.jpg *License*: unknown *Contributors*: Adziura, Antosh, BurgererSF, RaNo, Shalom Alechem, Verica Atrebatum, 3 anonymous edits

Image:Vytautas the great.jpg *Source*: http://en.wikipedia.org/w/index.php?title=File:Vytautas_the_great.jpg *License*: unknown *Contributors*: User:Mareczko

Image:Pieczęć Jagiełły.jpg *Source*: http://en.wikipedia.org/w/index.php?title=File:Pieczęć_Jagiełły.jpg *License*: unknown *Contributors*: User:Lestath

Image:Wladislaus II Jagiello of Poland.PNG *Source*: http://en.wikipedia.org/w/index.php?title=File:Wladislaus_II_Jagiello_of_Poland.PNG *License*: unknown *Contributors*: Azgar, BurgererSF, Kürschner, Mathiasrex

Image:Grunwald bitwa.jpg *Source*: http://en.wikipedia.org/w/index.php?title=File:Grunwald_bitwa.jpg *License*: unknown *Contributors*: AndreasPraefcke, BurgererSF, Czalex, Goldfritha, Hiuppo, Kaganer, Kevyn, Kirill Lokshin, Ludmiła Pilecka, M.K., Mathiasrex, Niki K, Perhelion, Piotrus, Skim, TOR, Wames, Wikipedysta300, 2 anonymous edits

File:Panorama of Malbork Castle, part 4.jpg *Source*: http://en.wikipedia.org/w/index.php?title=File:Panorama_of_Malbork_Castle,_part_4.jpg *License*: unknown *Contributors*: User:Carschten, User:DerHexer

Image:Polish and Lithuanian Conflict with Prussia. 1377-1435..png *Source*: http://en.wikipedia.org/w/index.php?title=File:Polish_and_Lithuanian_Conflict_with_Prussia._1377-1435..png *License*: unknown *Contributors*: Arz, GiW, Qp10qp

File:Flag_of_Poland.svg *Source*: http://en.wikipedia.org/w/index.php?title=File:Flag_of_Poland.svg *License*: unknown *Contributors*: Anomie, Mifter

File:Herb Polski.svg *Source*: http://en.wikipedia.org/w/index.php?title=File:Herb_Polski.svg *License*: unknown *Contributors*: Aotearoa, Denelson83, Frombenny, Gustavo Szwedowski de Korwin, Halibutt, Klemen Kocjancic, Madden, Odder, Zscout370, 3 anonymous edits

File:EU-Poland.svg *Source*: http://en.wikipedia.org/w/index.php?title=File:EU-Poland.svg *License*: unknown *Contributors*: User:NuclearVacuum

File:Increase2.svg *Source*: http://en.wikipedia.org/w/index.php?title=File:Increase2.svg *License*: unknown *Contributors*: Sarang

Image:Speakerlink.svg *Source*: http://en.wikipedia.org/w/index.php?title=File:Speakerlink.svg *License*: unknown *Contributors*: Woodstone. Original uploader was Woodstone at en.wikipedia

File:Polska 960 - 992.svg *Source*: http://en.wikipedia.org/w/index.php?title=File:Polska_960_-_992.svg *License*: unknown *Contributors*: User:Waćpan, user:Poznaniak

File:Zaprowadzenie chrzescijanstwa 965 Matejko.JPG *Source*: http://en.wikipedia.org/w/index.php?title=File:Zaprowadzenie_chrzescijanstwa_965_Matejko.JPG *License*: unknown *Contributors*: BurgererSF, Goldfritha, Kjetil r, Ludmiła Pilecka, Piotrus, Shakko, Wst, 3 anonymous edits

File:Jan Matejko-Astronomer Copernicus-Conversation with God.jpg *Source*: http://en.wikipedia.org/w/index.php?title=File:Jan_Matejko-Astronomer_Copernicus-Conversation_with_God.jpg *License*: unknown *Contributors*: Alaniaris, BurgererSF, Dirk Hünniger, EugeneZelenko, Gnesener1900, Goldfritha, Kürschner, Ludmiła Pilecka, Matthead, Olivier2, Piotrus, Pko, Plindenbaum, Slomox, Staszek Lem, Wames, Wst, 5 anonymous edits

File:Kraków - Wawel from Kopiec Krakusa.jpg *Source*: http://en.wikipedia.org/w/index.php?title=File:Kraków_-_Wawel_from_Kopiec_Krakusa.jpg *License*: unknown *Contributors*: User:RaNo

File:Rzeczpospolita 1600.png *Source*: http://en.wikipedia.org/w/index.php?title=File:Rzeczpospolita_1600.png *License*: unknown *Contributors*: Bandar Lego, Dcoetzee, Denniss, Electionworld, It Is Me Here

File:SiemiginowskiJerzy.1686.JanIIISobieskiPodWiedniem.jpg *Source*: http://en.wikipedia.org/w/index.php?title=File:SiemiginowskiJerzy.1686.JanIIISobieskiPodWiedniem.jpg *License*: unknown *Contributors*: AnonMoos, Boo-Boo Baroo, BurgererSF, Gryffindor, Infrogmation, Piotrus, Rotational, Shalom Alechem, Umherirrender, Wames, 1 anonymous edits

File:Stanisław II August Poniatowski in coronation clothes.PNG *Source*: http://en.wikipedia.org/w/index.php?title=File:Stanisław_II_August_Poniatowski_in_coronation_clothes.PNG *License*: unknown *Contributors*: Boo-Boo Baroo, BurgererSF, Darwinius, Ecummenic, Kürschner, Mathiasrex

File:Smuglewicz Kosciuszko 2.jpg *Source*: http://en.wikipedia.org/w/index.php?title=File:Smuglewicz_Kosciuszko_2.jpg *License*: unknown *Contributors*: Alma Pater, Bogomolov.PL, Cutwind, DieBuche, DocentX, Goldfritha, Man vyi, MarcusBritish, Mathiasrex, Sfu, Taxiarchos228, Wames, 8 anonymous edits

File:Starcie belwederczykow z kirasjerami rosyjskimi na moscie w Lazienkach.jpg *Source*: http://en.wikipedia.org/w/index.php?title=File:Starcie_belwederczykow_z_kirasjerami_rosyjskimi_na_moscie_w_Lazienkach.jpg *License*: unknown *Contributors*: Allen3, Bedford, Emax, Magog the Ogre, Piotrus, Raven4x4x

File:Jozef Pilsudski.jpg *Source*: http://en.wikipedia.org/w/index.php?title=File:Jozef_Pilsudski.jpg *License*: unknown *Contributors*: Alma Pater, Ejdzej, EugeneZelenko, Kocio, Mathiasrex, Serdelll, 1 anonymous edits

File:Rzeczpospolita 1939.svg *Source*: http://en.wikipedia.org/w/index.php?title=File:Rzeczpospolita_1939.svg *License*: unknown *Contributors*: User:Halibutt

File:10Dyw.JPG *Source*: http://en.wikipedia.org/w/index.php?title=File:10Dyw.JPG *License*: unknown *Contributors*: Andros64, GeorgHH, Jarekt, Mathiasrex, PMG, Peter Weis, Pibwl, Polish29, Robek, Sandmann4u, SuperTank17, Thib Phil, 3 anonymous edits

File:Curzon line en.svg *Source*: http://en.wikipedia.org/w/index.php?title=File:Curzon_line_en.svg *License*: unknown *Contributors*: User:radek.s

File:Polish Soldier's Grave Warsaw 1945.jpg *Source*: http://en.wikipedia.org/w/index.php?title=File:Polish_Soldier's_Grave_Warsaw_1945.jpg *License*: unknown *Contributors*: Tadeusz Bukowski

File:Solidarity poster 1989.jpg *Source*: http://en.wikipedia.org/w/index.php?title=File:Solidarity_poster_1989.jpg *License*: unknown *Contributors*: Andy Dingley, BurgererSF

File:Bundesarchiv Bild 183-F0417-0001-011, Berlin, VII. SED-Parteitag, Eröffnung.jpg *Source*: http://en.wikipedia.org/w/index.php?title=File:Bundesarchiv_Bild_183-F0417-0001-011,_Berlin,_VII._SED-Parteitag,_Eröffnung.jpg *License*: unknown *Contributors*: Fredy.00, Martin H., Mbdortmund, Trust Is All You Need

File:Flaga RP z UE.jpg *Source*: http://en.wikipedia.org/w/index.php?title=File:Flaga_RP_z_UE.jpg *License*: unknown *Contributors*: Michal Osmenda

File:Poland topo.jpg *Source*: http://en.wikipedia.org/w/index.php?title=File:Poland_topo.jpg *License*: unknown *Contributors*: Captain Blood, Hystrix, Juetho, Piotrus

File:Śląskie Kamienie - Dívčí Kameny.JPG *Source*: http://en.wikipedia.org/w/index.php?title=File:Śląskie_Kamienie_-_Dívčí_Kameny.JPG *License*: unknown *Contributors*: User:Pudelek

File:Giewont z Zakopanego.jpg *Source*: http://en.wikipedia.org/w/index.php?title=File:Giewont_z_Zakopanego.jpg *License*: unknown *Contributors*: User:Op

File:2 SPN 01.jpg *Source*: http://en.wikipedia.org/w/index.php?title=File:2_SPN_01.jpg *License*: unknown *Contributors*: BLueFiSH.as, Merlin, Panek, Piotrus, Reytan, Shalom Alechem, Uroboros, 2 anonymous edits

File:Modlin spichlerz.jpg *Source*: http://en.wikipedia.org/w/index.php?title=File:Modlin_spichlerz.jpg *License*: unknown *Contributors*: User:Wojsyl

File:Kurtkowiec i czerwone.jpg *Source*: http://en.wikipedia.org/w/index.php?title=File:Kurtkowiec_i_czerwone.jpg *License*: unknown *Contributors*: Adi, Adziura, Cathy Richards, Daniel Baránek, Dome, Michau Sm, Panek, Selso, Shalom Alechem, ToSter, Ulamm, 5 anonymous edits

File:Solina Poland.jpg *Source*: http://en.wikipedia.org/w/index.php?title=File:Solina_Poland.jpg *License*: unknown *Contributors*: Man vyi, Segu, Shalom Alechem, Silar, 2 anonymous edits

File:Zatoka Pucka - Bay of Puck (8).jpg *Source*: http://en.wikipedia.org/w/index.php?title=File:Zatoka_Pucka_-_Bay_of_Puck_(8).jpg *License*: unknown *Contributors*: Krzysztof, Silar, 1 anonymous edits

File:Biała Ręka Ojcowski PN.jpg *Source*: http://en.wikipedia.org/w/index.php?title=File:Biała_Ręka_Ojcowski_PN.jpg *License*: unknown *Contributors*: Jan Jerszyński

File:WhiteStorkFamily.jpg *Source*: http://en.wikipedia.org/w/index.php?title=File:WhiteStorkFamily.jpg *License*: unknown *Contributors*: User:Manfred Heyde

File:Wisent.jpg *Source*: http://en.wikipedia.org/w/index.php?title=File:Wisent.jpg *License*: unknown *Contributors*: Henryk Kotowski,

File:Plaża w Pucku - kitesurfers - beach in Puck (4).jpg *Source*: http://en.wikipedia.org/w/index.php?title=File:Plaża_w_Pucku_-_kitesurfers_-_beach_in_Puck_(4).jpg *License*: unknown *Contributors*: Fagairolles 34, GiW, Krzysztof, Marek Banach, 3 anonymous edits

File:Bronisław Komorowski rotcropped.jpg *Source*: http://en.wikipedia.org/w/index.php?title=File:Bronisław_Komorowski_rotcropped.jpg *License*: unknown *Contributors*: User:Saibo

File:PL Sejm hall.jpg *Source*: http://en.wikipedia.org/w/index.php?title=File:PL_Sejm_hall.jpg *License*: unknown *Contributors*: User:Network.nt

File:Polishsupremecourt.JPEG *Source*: http://en.wikipedia.org/w/index.php?title=File:Polishsupremecourt.JPEG *License*: unknown *Contributors*: Original uploader was Cjs2111 at en.wikipedia

File:Manuscript of the Constitution of the 3rd May 1791.PNG *Source*: http://en.wikipedia.org/w/index.php?title=File:Manuscript_of_the_Constitution_of_the_3rd_May_1791.PNG *License*: unknown *Contributors*: Ignacy Potocki (1750–1809), Hugo Kołłątaj (1750–1812), Stanisław August Poniatowski (1732–1798)

File:Radoslaw Sikorski meets Secretary Hillary Clinton.jpg *Source*: http://en.wikipedia.org/w/index.php?title=File:Radoslaw_Sikorski_meets_Secretary_Hillary_Clinton.jpg *License*: unknown *Contributors*: Michael Gross, State Department photographer

File:Walesa and Tusk EPP.jpg *Source*: http://en.wikipedia.org/w/index.php?title=File:Walesa_and_Tusk_EPP.jpg *License*: unknown *Contributors*: europeanpeoplesparty

Image:POL location map.svg *Source*: http://en.wikipedia.org/w/index.php?title=File:POL_location_map.svg *License*: unknown *Contributors*: Hiuppo

File:POL województwo zachodniopomorskie COA.svg *Source*: http://en.wikipedia.org/w/index.php?title=File:POL_województwo_zachodniopomorskie_COA.svg *License*: unknown *Contributors*: User:Bastianow

File:POL województwo pomorskie COA.svg *Source*: http://en.wikipedia.org/w/index.php?title=File:POL_województwo_pomorskie_COA.svg *License*: unknown *Contributors*: User:Bastianow

File:Warminsko-mazurskie herb.svg *Source*: http://en.wikipedia.org/w/index.php?title=File:Warminsko-mazurskie_herb.svg *License*: unknown *Contributors*: Derbeth, Frombenny, Gustavo Szwedowski de Korwin, JDavid, MAC13, Serdelll, 1 anonymous edits

File:POL województwo podlaskie COA.svg *Source*: http://en.wikipedia.org/w/index.php?title=File:POL_województwo_podlaskie_COA.svg *License*: unknown *Contributors*: Bastianow, Derbeth, Frombenny, Gustavo Szwedowski de Korwin, JDavid, WarX, XtraVert, 2 anonymous edits

File:POL województwo lubuskie COA.svg *Source*: http://en.wikipedia.org/w/index.php?title=File:POL_województwo_lubuskie_COA.svg *License*: unknown *Contributors*: User:Bastianow

File:POL województwo wielkopolskie COA.svg *Source*: http://en.wikipedia.org/w/index.php?title=File:POL_województwo_wielkopolskie_COA.svg *License*: unknown *Contributors*: User:Bastianow

File:POL województwo kujawsko-pomorskie COA.svg *Source*: http://en.wikipedia.org/w/index.php?title=File:POL_województwo_kujawsko-pomorskie_COA.svg *License*: unknown *Contributors*: Lech Tadeusz Karczewski

File:POL województwo dolnośląskie COA.svg *Source*: http://en.wikipedia.org/w/index.php?title=File:POL_województwo_dolnośląskie_COA.svg *License*: unknown *Contributors*: User:Poznaniak

File:POL województwo opolskie COA.svg *Source*: http://en.wikipedia.org/w/index.php?title=File:POL_województwo_opolskie_COA.svg *License*: unknown *Contributors*: user:Poznaniak

File:POL_województwo_śląskie_COA.svg *Source*: http://en.wikipedia.org/w/index.php?title=File:POL_województwo_śląskie_COA.svg *License*: unknown *Contributors*: user:Poznaniak

File:POL wojewodztwo świętokrzyskie COA.svg *Source*: http://en.wikipedia.org/w/index.php?title=File:POL_wojewodztwo_świętokrzyskie_COA.svg *License*: unknown *Contributors*: -

File:POL województwo łódzkie COA.svg *Source*: http://en.wikipedia.org/w/index.php?title=File:POL_województwo_łódzkie_COA.svg *License*: unknown *Contributors*: User:Bastianow

File:POL województwo mazowieckie COA.svg *Source*: http://en.wikipedia.org/w/index.php?title=File:POL_województwo_mazowieckie_COA.svg *License*: unknown *Contributors*: User:Poznaniak

File:POL województwo lubelskie COA.svg *Source*: http://en.wikipedia.org/w/index.php?title=File:POL_województwo_lubelskie_COA.svg *License*: unknown *Contributors*: user:MesserWoland

File:POL województwo małopolskie COA.svg *Source*: http://en.wikipedia.org/w/index.php?title=File:POL_województwo_małopolskie_COA.svg *License*: unknown *Contributors*: user:Poznaniak

File:POL województwo podkarpackie COA.svg *Source*: http://en.wikipedia.org/w/index.php?title=File:POL_województwo_podkarpackie_COA.svg *License*: unknown *Contributors*: Derbeth, Frombenny, Gustavo Szwedowski de Korwin, Jaron11, Mir, Valentinian

File:Krzesiny 8RB.JPG *Source*: http://en.wikipedia.org/w/index.php?title=File:Krzesiny_8RB.JPG *License*: unknown *Contributors*: User:Radomil

File:Polish Army soldiers in Afghanistan.jpg *Source*: http://en.wikipedia.org/w/index.php?title=File:Polish_Army_soldiers_in_Afghanistan.jpg *License*: unknown *Contributors*: Ministerstwo Obrony Narodowej

File:Central Silesian Park - Mounted police 02.jpg *Source*: http://en.wikipedia.org/w/index.php?title=File:Central_Silesian_Park_-_Mounted_police_02.jpg *License*: unknown *Contributors*: User:Yarl

File:Warsaw7ob.jpg *Source*: http://en.wikipedia.org/w/index.php?title=File:Warsaw7ob.jpg *License*: unknown *Contributors*: User:DocentX

File:Schengenzone.svg *Source*: http://en.wikipedia.org/w/index.php?title=File:Schengenzone.svg *License*: unknown *Contributors*: User:CrazyPhunk

File:POL.Gdynia.SeaTowers.DarPomorza2009.jpg *Source*: http://en.wikipedia.org/w/index.php?title=File:POL.Gdynia.SeaTowers.DarPomorza2009.jpg *License*: unknown *Contributors*: User:Wames

File:Gielda3 Wwa beax.jpg *Source*: http://en.wikipedia.org/w/index.php?title=File:Gielda3_Wwa_beax.jpg *License*: unknown *Contributors*: User:Beax

File:Warsaw Downtown.jpg *Source*: http://en.wikipedia.org/w/index.php?title=File:Warsaw_Downtown.jpg *License*: unknown *Contributors*: User:DocenttX

File:Ratusz Zamosc.jpg *Source*: http://en.wikipedia.org/w/index.php?title=File:Ratusz_Zamosc.jpg *License*: unknown *Contributors*: Original uploader was Mceurytos at pl.wikipedia

File:Wrocław, Iglica a Hala století.jpg *Source*: http://en.wikipedia.org/w/index.php?title=File:Wrocław,_Iglica_a_Hala_století.jpg *License*: unknown *Contributors*: Aktron

File:Elektrownia Zeran.jpg *Source*: http://en.wikipedia.org/w/index.php?title=File:Elektrownia_Zeran.jpg *License*: unknown *Contributors*: Original uploader was Hiuppo at pl.wikipedia

File:SP-LRA 2.png *Source*: http://en.wikipedia.org/w/index.php?title=File:SP-LRA_2.png *License*: unknown *Contributors*: © Strower

File:A4 Krakow 20070815 1026.jpg *Source*: http://en.wikipedia.org/w/index.php?title=File:A4_Krakow_20070815_1026.jpg *License*: unknown *Contributors*: User:Jakubhal

File:Marie Curie (Nobel-Chem).png *Source*: http://en.wikipedia.org/w/index.php?title=File:Marie_Curie_(Nobel-Chem).png *License*: unknown *Contributors*: Afrank99, Bohème, DEDB, Fastfission, Kjunix, Martin H., Materialscientist, 3 anonymous edits

File:Warsaw Easter 2008 028.jpg *Source*: http://en.wikipedia.org/w/index.php?title=File:Warsaw_Easter_2008_028.jpg *License*: unknown *Contributors*: Wizzard

File:Poczta Główna Bydg od Brdy a.jpg *Source*: http://en.wikipedia.org/w/index.php?title=File:Poczta_Główna_Bydg_od_Brdy_a.jpg *License*: unknown *Contributors*: User:Pit1233

File:TPSA tower warsaw.JPG *Source*: http://en.wikipedia.org/w/index.php?title=File:TPSA_tower_warsaw.JPG *License*: unknown *Contributors*: User:Masti

File:Gdańsk kamienice przy Długim Targu.jpg *Source*: http://en.wikipedia.org/w/index.php?title=File:Gdańsk_kamienice_przy_Długim_Targu.jpg *License*: unknown *Contributors*: User:Sfu

File:Krakau Markt.jpg *Source*: http://en.wikipedia.org/w/index.php?title=File:Krakau_Markt.jpg *License*: unknown *Contributors*: User:Diether

File:Lichen stary Poland.jpg *Source*: http://en.wikipedia.org/w/index.php?title=File:Lichen_stary_Poland.jpg *License*: unknown *Contributors*: User:Czechu81

File:CerkiewStDuchaBialystok.JPG *Source*: http://en.wikipedia.org/w/index.php?title=File:CerkiewStDuchaBialystok.JPG *License*: unknown *Contributors*: Akir, Ludmiła Pilecka, NVO, Przykuta, Roman Miller, Wouterhagens, Wp, 2 anonymous edits

File:Krakow Synagoga Tempel 20071111 1123 2077.jpg *Source*: http://en.wikipedia.org/w/index.php?title=File:Krakow_Synagoga_Tempel_20071111_1123_2077.jpg *License*: unknown *Contributors*: User:Jakubhal

File:MUG Hospital Bldg 1.jpg *Source*: http://en.wikipedia.org/w/index.php?title=File:MUG_Hospital_Bldg_1.jpg *License*: unknown *Contributors*: User:JohnnyTremolo

File:Collegium Maius 07.JPG *Source*: http://en.wikipedia.org/w/index.php?title=File:Collegium_Maius_07.JPG *License*: unknown *Contributors*: User:Cancre

File:Collegium Minus - UAM.jpg *Source*: http://en.wikipedia.org/w/index.php?title=File:Collegium_Minus_-_UAM.jpg *License*: unknown *Contributors*: User:Mains

File:Ryszard Kaczorowski 21 02 2008 (4).jpg *Source*: http://en.wikipedia.org/w/index.php?title=File:Ryszard_Kaczorowski_21_02_2008_(4).jpg *License*: unknown *Contributors*: Archiwum Kancelarii Prezydenta RP

File:UniwersytetWrocławski.jpg *Source*: http://en.wikipedia.org/w/index.php?title=File:UniwersytetWrocławski.jpg *License*: unknown *Contributors*: User:JacobJ

File:Nikolaus Kopernikus.jpg *Source*: http://en.wikipedia.org/w/index.php?title=File:Nikolaus_Kopernikus.jpg *License*: unknown *Contributors*: ArséniureDeGallium, ArtMechanic, Ausir, Craigboy, Editor at Large, J.delanoy, Kyro, Manuelt15, Matthead, Mikkalai, Pko, Samuel Grant, TarmoK, ThomasPusch, 6 anonymous edits

File:Grand Hotel Sopot.JPG *Source*: http://en.wikipedia.org/w/index.php?title=File:Grand_Hotel_Sopot.JPG *License*: unknown *Contributors*: Alina Zienowicz, e-mail

File:John Paul II Medal of Freedom 2004.jpg *Source*: http://en.wikipedia.org/w/index.php?title=File:John_Paul_II_Medal_of_Freedom_2004.jpg *License*: unknown *Contributors*: Eric Draper

File:Frederic Chopin photo.jpeg *Source*: http://en.wikipedia.org/w/index.php?title=File:Frederic_Chopin_photo.jpeg *License*: unknown *Contributors*: Louis-Auguste Bisson, very old and poor copy, completely restored and remastered by Amano1

File:Tomasz Stanko-CP.jpg *Source*: http://en.wikipedia.org/w/index.php?title=File:Tomasz_Stanko-CP.jpg *License*: unknown *Contributors*: Cezary Piwowarski

File:Fakty TVP-studio Wroclaw 2008.jpg *Source*: http://en.wikipedia.org/w/index.php?title=File:Fakty_TVP-studio_Wroclaw_2008.jpg *License*: unknown *Contributors*: Original uploader was SNOL at pl.wikipedia

File:Adam Mickiewicz.PNG *Source*: http://en.wikipedia.org/w/index.php?title=File:Adam_Mickiewicz.PNG *License*: unknown *Contributors*: Arek1979, Bohème, BurgererSF, Butko, Darwinius, Freta, Kilom691, Krzysiu Jarzyna, Mathiasrex, Mattes, Rocket000, Shakko, Thuresson, TwoWings, 2 anonymous edits

File:Henryk Sienkiewicz.PNG *Source*: http://en.wikipedia.org/w/index.php?title=File:Henryk_Sienkiewicz.PNG *License*: unknown *Contributors*: Kazimierz Mordasewicz

File:Władysław Reymont.jpg *Source*: http://en.wikipedia.org/w/index.php?title=File:Władysław_Reymont.jpg *License*: unknown *Contributors*: Andros64, Demut, HuBar, 2 anonymous edits

File:Witold Gombrowicz by Bohdan Paczowski - detail.jpg *Source*: http://en.wikipedia.org/w/index.php?title=File:Witold_Gombrowicz_by_Bohdan_Paczowski_-_detail.jpg *License*: unknown *Contributors*: Bohdan_Paczowski

File:Czeslaw Milosz, 1986.jpg *Source*: http://en.wikipedia.org/w/index.php?title=File:Czeslaw_Milosz,_1986.jpg *License*: unknown *Contributors*: User:MDCarchives

File:Ratusz2007.jpg *Source*: http://en.wikipedia.org/w/index.php?title=File:Ratusz2007.jpg *License*: unknown *Contributors*: adamchemik

File:Kościół parafialny p.w. Wniebowzięcia NMP (Mariacki), Kraków, Rynek Główny, A-3 01.jpg *Source*: http://en.wikipedia.org/w/index.php?title=File:Kościół_parafialny_p.w._Wniebowzięcia_NMP_(Mariacki),_Kraków,_Rynek_Główny,_A-3_01.jpg *License*: unknown *Contributors*: User:Piotrekwas

File:Kotlet schabowy.jpg *Source*: http://en.wikipedia.org/w/index.php?title=File:Kotlet_schabowy.jpg *License*: unknown *Contributors*: phototram

File:Openstreetmap logo.svg *Source*: http://en.wikipedia.org/w/index.php?title=File:Openstreetmap_logo.svg *License*: unknown *Contributors*: OpenStreetMap

File:6 Lask 5.jpg *Source*: http://en.wikipedia.org/w/index.php?title=File:6_Lask_5.jpg *License*: unknown *Contributors*: Jarekt, Shalom Alechem, 3 anonymous edits

File:POL Łask COA.svg *Source*: http://en.wikipedia.org/w/index.php?title=File:POL_Łask_COA.svg *License*: unknown *Contributors*: Frombenny, Masur, Orem, Raf24, Umherirrender, 1 anonymous edits

File:POL powiat łaski flag.svg *Source*: http://en.wikipedia.org/w/index.php?title=File:POL_powiat_łaski_flag.svg *License*: unknown *Contributors*: Bastianow, Denelson83, Homo lupus, Tiikeri, 1 anonymous edits

File:POL powiat łaski COA.svg *Source*: http://en.wikipedia.org/w/index.php?title=File:POL_powiat_łaski_COA.svg *License*: unknown *Contributors*: Bastianow, Gustavo Szwedowski de Korwin, Yarl, 1 anonymous edits

File:POL_województwo_łódzkie_powiat_łaski_map.svg *Source*: http://en.wikipedia.org/w/index.php?title=File:POL_województwo_łódzkie_powiat_łaski_map.svg *License*: unknown *Contributors*: user:MariuszR, user:WarX

File:Powiat_łaski.svg *Source*: http://en.wikipedia.org/w/index.php?title=File:Powiat_łaski.svg *License*: unknown *Contributors*: User:JaqbS

File:POL gmina Widawa COA.svg *Source*: http://en.wikipedia.org/w/index.php?title=File:POL_gmina_Widawa_COA.svg *License*: unknown *Contributors*: User:Semu

File:Rougon Alpes de Haute Provence France.jpg *Source*: http://en.wikipedia.org/w/index.php?title=File:Rougon_Alpes_de_Haute_Provence_France.jpg *License*: unknown *Contributors*: User:JialiangGao

File:KippelLötschental WoodenHouses.jpg *Source*: http://en.wikipedia.org/w/index.php?title=File:KippelLötschental_WoodenHouses.jpg *License*: unknown *Contributors*: Herzi Pinki

File:Ourika berbere village.jpg *Source*: http://en.wikipedia.org/w/index.php?title=File:Ourika_berbere_village.jpg *License*: unknown *Contributors*: Fathzer / Jean-Marc Astesana from Bretonneux, France

File:Guri Rajasthan 02.jpg *Source*: http://en.wikipedia.org/w/index.php?title=File:Guri_Rajasthan_02.jpg *License*: unknown *Contributors*: User:Entomolo

File:View of RTallavalasa.jpg *Source*: http://en.wikipedia.org/w/index.php?title=File:View_of_RTallavalasa.jpg *License*: unknown *Contributors*: User:Adityamadhav83

File:Ogi Shirakawa02bs3200.jpg *Source*: http://en.wikipedia.org/w/index.php?title=File:Ogi_Shirakawa02bs3200.jpg *License*: unknown *Contributors*: 663highland

File:Pariangan.jpg *Source*: http://en.wikipedia.org/w/index.php?title=File:Pariangan.jpg *License*: unknown *Contributors*: User:MichaelJLowe

File:Kovachevitsa.jpg *Source*: http://en.wikipedia.org/w/index.php?title=File:Kovachevitsa.jpg *License*: unknown *Contributors*: User:Katieleu

File:Village near lake baikal 45230241 f3b18ea67c b.jpg *Source*: http://en.wikipedia.org/w/index.php?title=File:Village_near_lake_baikal_45230241_f3b18ea67c_b.jpg *License*: unknown *Contributors*: Nagyman (flickr)

File:Kosmach Doroha.jpg *Source*: http://en.wikipedia.org/w/index.php?title=File:Kosmach_Doroha.jpg *License*: unknown *Contributors*: User:Yakudza

File:castle.combe.mainstreet.arp.jpg *Source*: http://en.wikipedia.org/w/index.php?title=File:Castle.combe.mainstreet.arp.jpg *License*: unknown *Contributors*: User:Arpingstone

File:Bisley.jpg *Source*: http://en.wikipedia.org/w/index.php?title=File:Bisley.jpg *License*: unknown *Contributors*: User:Jongleur100

File:Saint-Cirq-Lapopie.jpg *Source*: http://en.wikipedia.org/w/index.php?title=File:Saint-Cirq-Lapopie.jpg *License*: unknown *Contributors*: Adam Baker from Houston / Moscow / Toulouse (travel a lot)

File:saifivillage.JPG *Source*: http://en.wikipedia.org/w/index.php?title=File:Saifivillage.JPG *License*: unknown *Contributors*: Elie plus, Linaduliban

File:General view to Al annaze1.jpg *Source*: http://en.wikipedia.org/w/index.php?title=File:General_view_to_Al_annaze1.jpg *License*: unknown *Contributors*: User:Same7same7

File:Puamau.jpg *Source*: http://en.wikipedia.org/w/index.php?title=File:Puamau.jpg *License*: unknown *Contributors*: American at de.wikipedia

File:Carlb-fogo-newfoundland-2002.jpg *Source*: http://en.wikipedia.org/w/index.php?title=File:Carlb-fogo-newfoundland-2002.jpg *License*: unknown *Contributors*: Conscious, Croquant, EvaK, Peregrine981, Telim tor, WayneRay, 1 anonymous edits

File:Newfane vermont fall 2009.JPG *Source*: http://en.wikipedia.org/w/index.php?title=File:Newfane_vermont_fall_2009.JPG *License*: unknown *Contributors*: chensiyuan

File:A Village in Kaita.jpg *Source*: http://en.wikipedia.org/w/index.php?title=File:A_Village_in_Kaita.jpg *License*: unknown *Contributors*: User:Shamsurabiu

GNU Free Documentation License Version 1.2, November 2002 Copyright (C) 2000,2001,2002 Free Software Foundation, Inc. 59 Temple Place, Suite 330, Boston, MA 02111-1307 USA Everyone is permitted to copy and distribute verbatim copies of this license document, but changing it is not allowed.

0. PREAMBLE

The purpose of this License is to make a manual, textbook, or other functional and useful document "free" in the sense of freedom: to assure everyone the effective freedom to copy and redistribute it, with or without modifying it, either commercially or noncommercially. Secondarily, this License preserves for the author and publisher a way to get credit for their work, while not being considered responsible for modifications made by others. This License is a kind of "copyleft", which means that derivative works of the document must themselves be free in the same sense. It complements the GNU General Public License, which is a copyleft license designed for free software. We have designed this License in order to use it for manuals for free software, because free software needs free documentation: a free program should come with manuals providing the same freedoms that the software does. But this License is not limited to software manuals; it can be used for any textual work, regardless of subject matter or whether it is published as a printed book. We recommend this License principally for works whose purpose is instruction or reference.

1. APPLICABILITY AND DEFINITIONS

This License applies to any manual or other work, in any medium, that contains a notice placed by the copyright holder saying it can be distributed under the terms of this License. Such a notice grants a world-wide, royalty-free license, unlimited in duration, to use that work under the conditions stated herein. The "Document", below, refers to any such manual or work. Any member of the public is a licensee, and is addressed as "you". You accept the license if you copy, modify or distribute the work in a way requiring permission under copyright law. A "Modified Version" of the Document means any work containing the Document or a portion of it, either copied verbatim, or with modifications and/or translated into another language. A "Secondary Section" is a named appendix or a front-matter section of the Document that deals exclusively with the relationship of the publishers or authors of the Document to the Document's overall subject (or to related matters) and contains nothing that could fall directly within that overall subject. (Thus, if the Document is in part a textbook of mathematics, a Secondary Section may not explain any mathematics.) The relationship could be a matter of historical connection with the subject or with related matters, or of legal, commercial, philosophical, ethical or political position regarding them. The "Invariant Sections" are certain Secondary Sections whose titles are designated, as being those of Invariant Sections, in the notice that says that the Document is released under this License. If a section does not fit the above definition of Secondary then it is not allowed to be designated as Invariant. The Document may contain zero Invariant Sections. If the Document does not identify any Invariant Sections then there are none. The "Cover Texts" are certain short passages of text that are listed, as Front-Cover Texts or Back-Cover Texts, in the notice that says that the Document is released under this License. A Front-Cover Text may be at most 5 words, and a Back-Cover Text may be at most 25 words. A "Transparent" copy of the Document means a machine-readable copy, represented in a format whose specification is available to the general public, that is suitable for revising the document straightforwardly with generic text editors or (for images composed of pixels) generic paint programs or (for drawings) some widely available drawing editor, and that is suitable for input to text formatters or for automatic translation to a variety of formats suitable for input to text formatters. A copy made in an otherwise Transparent file format whose markup, or absence of markup, has been arranged to thwart or discourage subsequent modification by readers is not Transparent. An image format is not Transparent if used for any substantial amount of text. A copy that is not "Transparent" is called "Opaque". Examples of suitable formats for Transparent copies include plain ASCII without markup, Texinfo input format, LaTeX input format, SGML or XML using a publicly available DTD, and standard-conforming simple HTML, PostScript or PDF designed for human modification. Examples of transparent image formats include PNG, XCF and JPG. Opaque formats include proprietary formats that can be read and edited only by proprietary word processors, SGML or XML for which the DTD and/or processing tools are not generally available, and the machine-generated HTML, PostScript or PDF produced by some word processors for output purposes only. The "Title Page" means, for a printed book, the title page itself, plus such following pages as are needed to hold, legibly, the material this License requires to appear in the title page. For works in formats which do not have any title page as such, "Title Page" means the text near the most prominent appearance of the work's title, preceding the beginning of the body of the text. A section "Entitled XYZ" means a named subunit of the Document whose title either is precisely XYZ or contains XYZ in parentheses following text that translates XYZ in another language. (Here XYZ stands for a specific section name mentioned below, such as "Acknowledgements", "Dedications", "Endorsements", or "History".) To "Preserve the Title" of such a section when you modify the Document means that it remains a section "Entitled XYZ" according to this definition. The Document may include Warranty Disclaimers next to the notice which states that this License applies to the Document. These Warranty Disclaimers are considered to be included by reference in this License, but only as regards disclaiming warranties: any other implication that these Warranty Disclaimers may have is void and has no effect on the meaning of this License.

2. VERBATIM COPYING

You may copy and distribute the Document in any medium, either commercially or noncommercially, provided that this License, the copyright notices, and the license notice saying this License applies to the Document are reproduced in all copies, and that you add no other conditions whatsoever to those of this License. You may not use technical measures to obstruct or control the reading or further copying of the copies you make or distribute. However, you may accept compensation in exchange for copies. If you distribute a large enough number of copies you must also follow the conditions in section 3. You may also lend copies, under the same conditions stated above, and you may publicly display copies.

3. COPYING IN QUANTITY

If you publish printed copies (or copies in media that commonly have printed covers) of the Document, numbering more than 100, and the Document's license notice requires Cover Texts, you must enclose the copies in covers that carry, clearly and legibly, all these Cover Texts: Front-Cover Texts on the front cover, and Back-Cover Texts on the back cover. Both covers must also clearly and legibly identify you as the publisher of these copies. The front cover must present the full title with all words of the title equally prominent and visible. You may add other material on the covers in addition. Copying with changes limited to the covers, as long as they preserve the title of the Document and satisfy these conditions, can be treated as verbatim copying in other respects. If the required texts for either cover are too voluminous to fit legibly, you should put the first ones listed (as many as fit reasonably) on the actual cover, and continue the rest onto adjacent pages. If you publish or distribute Opaque copies of the Document numbering more than 100, you must either include a machine-readable Transparent copy along with each Opaque copy, or state in or with each Opaque copy a computer-network location from which the general network-using public has access to download using public-standard network protocols a complete Transparent copy of the Document, free of added material. If you use the latter option, you must take reasonably prudent steps, when you begin distribution of Opaque copies in quantity, to ensure that this Transparent copy will remain thus accessible at the stated location until at least one year after the last time you distribute an Opaque copy (directly or through your agents or retailers) of that edition to the public. It is requested, but not required, that you contact the authors of the Document well before redistributing any large number of copies, to give them a chance to provide you with an updated version of the Document.

4. MODIFICATIONS

You may copy and distribute a Modified Version of the Document under the conditions of sections 2 and 3 above, provided that you release the Modified Version under precisely this License, with the Modified Version filling the role of the Document, thus licensing distribution and modification of the Modified Version to whoever possesses a copy of it. In addition, you must do these things in the Modified Version: A. Use in the Title Page (and on the covers, if any) a title distinct from that of the Document, and from those of previous versions (which should, if there were any, be listed in the History section of the Document). You may use the same title as a previous version if the original publisher of that version gives permission. B. List on the Title Page, as authors, one or more persons or entities responsible for authorship of the modifications in the Modified Version, together with at least five of the principal authors of the Document (all of its principal authors, if it has fewer than five), unless they release you from this requirement. C. State on the Title page the name of the publisher of the Modified Version, as the publisher. D. Preserve all the copyright notices of the Document. E. Add an appropriate copyright notice for your modifications adjacent to the other copyright notices. F. Include, immediately after the copyright notices, a license notice giving the public permission to use the Modified Version under the terms of this License, in the form shown in the Addendum below. G. Preserve in that license notice the full lists of Invariant Sections and required Cover Texts given in the Document's license notice. H. Include an unaltered copy of this License. I. Preserve the section Entitled "History", Preserve its Title, and add to it an item stating at least the title, year, new authors, and publisher of the Modified Version as given on the Title Page. If there is no section Entitled "History" in the Document, create one stating the title, year, authors, and publisher of the Document as given on its Title Page, then add an item describing the Modified Version as stated in the previous sentence. J. Preserve the network location, if any, given in the Document for public access to a Transparent copy of the Document, and likewise the network locations given in the Document for previous versions it was based on. These may be placed in the "History" section. You may omit a network location for a work that was published at least four years before the Document itself, or if the original publisher of the version it refers to gives permission. K. For any section Entitled "Acknowledgements" or "Dedications", Preserve the Title of the section, and preserve in the section all the substance and tone of each of the contributor acknowledgements and/or dedications given therein. L. Preserve all the Invariant Sections of the Document, unaltered in their text and in their titles. Section numbers or the equivalent are not considered part of the section titles. M. Delete any section Entitled "Endorsements". Such a section may not be included in the Modified Version. N. Do not retitle any existing section to be Entitled "Endorsements" or to conflict in title with any Invariant Section. O. Preserve any Warranty Disclaimers. If the Modified Version includes new front-matter sections or appendices that qualify as Secondary Sections and contain no material copied from the Document, you may at your option designate some or all of these sections as invariant. To do this, add their titles to the list of Invariant Sections in the Modified Version's license notice. These titles must be distinct from any other section titles. You may add a section Entitled "Endorsements", provided it contains nothing but endorsements of your Modified Version by various parties--for example, statements of peer review or that the text has been approved by an organization as the authoritative definition of a standard. You may add a passage of up to five words as a Front-Cover Text, and a passage of up to 25 words as a Back-Cover Text, to the end of the list of Cover Texts in the Modified Version. Only one passage of Front-Cover Text and one of Back-Cover Text may be added by (or through arrangements made by) any one entity. If the Document already includes a cover text for the same cover, previously added by you or by arrangement made by the same entity you are acting on behalf of, you may not add another; but you may replace the old one, on explicit permission from the previous publisher that added the old one. The author(s) and publisher(s) of the Document do not by this License give permission to use their names for publicity for or to assert or imply endorsement of any Modified Version.

5. COMBINING DOCUMENTS

You may combine the Document with other documents released under this License, under the terms defined in section 4 above for modified versions, provided that you include in the combination all of the Invariant Sections of all of the original documents, unmodified, and list them all as Invariant Sections of your combined work in its license notice, and that you preserve all their Warranty Disclaimers. The combined work need only contain one copy of this License, and multiple identical Invariant Sections may be replaced with a single copy. If there are multiple Invariant Sections with the same name but different contents, make the title of each such section unique by adding at the end of it, in parentheses, the name of the original author or publisher of that section if known, or else a unique number. Make the same adjustment to the section titles in the list of Invariant Sections in the license notice of the combined work. In the combination, you must combine any sections Entitled "History" in the various original documents, forming one section Entitled "History"; likewise combine any sections Entitled "Acknowledgements", and any sections Entitled "Dedications". You must delete all sections Entitled "Endorsements".

6. COLLECTIONS OF DOCUMENTS

You may make a collection consisting of the Document and other documents released under this License, and replace the individual copies of this License in the various documents with a single copy that is included in the collection, provided that you follow the rules of this License for verbatim copying of each of the documents in all other respects. You may extract a single document from such a collection, and distribute it individually under this License, provided you insert a copy of this License into the extracted document, and follow this License in all other respects regarding verbatim copying of that document.

7. AGGREGATION WITH INDEPENDENT WORKS

A compilation of the Document or its derivatives with other separate and independent documents or works, in or on a volume of a storage or distribution medium, is called an "aggregate" if the copyright resulting from the compilation is not used to limit the legal rights of the compilation's users beyond what the individual works permit. When the Document is included in an aggregate, this License does not apply to the other works in the aggregate which are not themselves derivative works of the Document. If the Cover Text requirement of section 3 is applicable to these copies of the Document, then if the Document is less than one half of the entire aggregate, the Document's Cover Texts may be placed on covers that bracket the Document within the aggregate, or the electronic equivalent of covers if the Document is in electronic form. Otherwise they must appear on printed covers that bracket the whole aggregate.

8. TRANSLATION

Translation is considered a kind of modification, so you may distribute translations of the Document under the terms of section 4. Replacing Invariant Sections with translations requires special permission from their copyright holders, but you may include translations of some or all Invariant Sections in addition to the original versions of these Invariant Sections. You may include a translation of this License, and all the license notices in the Document, and any Warranty Disclaimers, provided that you also include the original English version of this License and the original versions of those notices and disclaimers. In case of a disagreement between the translation and the original version of this License or a notice or disclaimer, the original version will prevail. If a section in the Document is Entitled "Acknowledgements", "Dedications", or "History", the requirement (section 4) to Preserve its Title (section 1) will typically require changing the actual title.

9. TERMINATION

You may not copy, modify, sublicense, or distribute the Document except as expressly provided for under this License. Any other attempt to copy, modify, sublicense or distribute the Document is void, and will automatically terminate your rights under this License. However, parties who have received copies, or rights, from you under this License will not have their licenses terminated so long as such parties remain in full compliance.

10. FUTURE REVISIONS OF THIS LICENSE

The Free Software Foundation may publish new, revised versions of the GNU Free Documentation License from time to time. Such new versions will be similar in spirit to the present version, but may differ in detail to address new problems or concerns. See http://www.gnu.org/copyleft/. Each version of the License is given a distinguishing version number. If the Document specifies that a particular numbered version of this License "or any later version" applies to it, you have the option of following the terms and conditions either of that specified version or of any later version that has been published (not as a draft) by the Free Software Foundation. If the Document does not specify a version number of this License, you may choose any version ever published (not as a draft) by the Free Software Foundation. ADDENDUM: How to use this License for your documents To use this License in a document you have written, include a copy of the License in the document and put the following copyright and license notices just after the title page: Copyright (c) YEAR YOUR NAME. Permission is granted to copy, distribute and/or modify this document under the terms of the GNU Free Documentation License, Version 1.2 or any later version published by the Free Software Foundation; with no Invariant Sections, no Front-Cover Texts, and no Back-Cover Texts. A copy of the license is included in the section entitled "GNU Free Documentation License". If you have Invariant Sections, Front-Cover Texts and Back-Cover Texts, replace the "with...Texts." line with this: with the Invariant Sections being LIST THEIR TITLES, with the Front-Cover Texts being LIST, and with the Back-Cover Texts being LIST. If you have Invariant Sections without Cover Texts, or some other combination of the three, merge those two alternatives to suit the situation. If your document contains nontrivial examples of program code, we recommend releasing these examples in parallel under your choice of free software license, such as the GNU General Public License, to permit their use in free software.

MIX
Papier aus verantwortungsvollen Quellen
Paper from responsible sources
FSC® C105338

Printed by Books on Demand GmbH, Norderstedt / Germany